GROUNDWATER VULNERABILITY ASSESSME

SELECTED PAPERS ON HYDROGEOLOGY

11

Series Editor: Dr. Nick S. Robins
Editor-in-Chief IAH Book Series
British Geological Survey
Wallingford, UK

INTERNATIONAL ASSOCIATION OF HYDROGEOLOGISTS

Groundwater Vulnerability Assessment and Mapping

Selected papers from the Groundwater Vulnerability Assessment and Mapping International Conference, Ustroń, Poland, 2004

Edited by

Andrzej J. Witkowski & Andrzej Kowalczyk
University of Silesia, Sosnowiec, Poland

Jaroslav Vrba
UNESCO Consultant and Chairman of IAH Commission on Groundwater Protection, Prague, Czech Republic

CRC Press
Taylor & Francis Group
Boca Raton London New York

CRC Press is an imprint of the
Taylor & Francis Group, an **informa** business
A TAYLOR & FRANCIS BOOK

CRC Press
Taylor & Francis Group
6000 Broken Sound Parkway NW, Suite 300
Boca Raton, FL 33487-2742

First issued in paperback 2019

© 2007 by Taylor & Francis Group, LLC
CRC Press is an imprint of Taylor & Francis Group, an Informa business

Typeset by MPS Limited, Chennai, India

No claim to original U.S. Government works

ISBN-13: 978-0-415-44561-0 (hbk)
ISBN-13: 978-0-367-38869-0 (pbk)

Library of Congress Cataloging-in-Publication Data

Groundwater Vulnerability Assessment and Mapping International Conference (2004 : Ustron, Poland)
    Groundwater vulnerability assessment and mapping : selected papers from the Groundwater Vulnerability Assessment and Mapping International Conference : Ustron, Poland, 2004 / edited by Andrzej J. Witkowski, Andrzej Kowalczyk and Jaroslav Vrba.
        p. cm.
    Includes bibliographical references and index.
    ISBN 978-0-415-44561-0 (hbk. : alk. paper) 1. Groundwater—Pollution—Congresses. 2. Water—Pollution potential—Congresses. 3. Water—Pollution potential—Computer simulation—Congresses. 4. Groundwater—Pollution–Computer simulation—Congresses. 5. Hydrogeology—Congresses. I. Witkowski, Andrzej, doc. dr. II. Kowalczyk, Andrzej. III. Vrba, J. IV. Title.

TD426.G775 2007
628.1'68—dc22   2007013657

Visit the Taylor & Francis Web site at
http://www.taylorandfrancis.com

and the CRC Press Web site at
http://www.crcpress.com

Table of contents

Preface

The papers in this volume cover the main issues and topics pertaining to the assessment of groundwater vulnerability:

- Factors affecting the selection of vulnerability assessment methods.
- Uncertainty in vulnerability assessment methodologies.
- Designing general (intrinsic) and specific vulnerability maps.
- Applications and limitations of vulnerability maps.
- Data needs, processing and presentation using a GIS format.

This volume has been divided into three main parts. The first part is a keynote introductory paper by Stephen Foster (the present IAH President). The second part covers aspects of the European approach to groundwater vulnerability assessment and mapping. Many of these papers refer to the Water Framework Directive (2000/60/EEC), which is currently being implemented by member states of the European Union, and a number of European funded programmes (e.g. COST Action 620), and some national solutions (Chapters 1–4). The third and most extensive part of this volume comprises case studies dealing with a wide range of issues concerning the assessment of intrinsic and specific vulnerability of different aquifers located in various geological and hydrogeological environments (coastal aquifers, shallow and deeper Quaternary aquifers, carbonate aquifers).

Increasing interest in the most vulnerable aquifers and specifically in karst aquifers has been confirmed by the large number of contributions on that topic. Because of the frequent application of the vulnerability assessment methodologies the case studies have been divided into two separate groups depending on the type of aquifer. The first group incorporates the porous-type aquifers (Chapters 5–15), whereas the second one deals with the karst-type aquifers (Chapters 16–22).

The papers were selected from those presented at the IAH Conference "Groundwater vulnerability assessment and mapping" which was held in Ustroń, Poland on 15–18 June 2004. The Conference was organized 10 years after the first book about groundwater vulnerability assessment and mapping was published by UNESCO and IAH. Within the fourth phase of the IHP the international working group, composed mainly of members of the IAH Groundwater Protection Commission, published the manual "Guidebook on Mapping Groundwater Vulnerability" (J. Vrba and A. Zaporozec, editors, 1994).

The Conference was organized by the IAH Commission on Groundwater Protection, the Polish National Chapter of IAH, UNESCO and the University of Silesia. A total of 81 papers were presented during the oral and poster sessions during the conference and 23 of them have been selected for inclusion in this volume. We would like to express our particularly warm acknowledgments to the authors for their contributions as well as for their patience and understanding with regard to the delay in publication which has been caused by editorial and technical reasons. We thank the numerous reviewers (Brian Adams, Alistar Allen, Colin Both, John Chilton, Antonio Cimino, Donal Daly, David Drew, Stephen Foster, Edmund Gosk, Ricardo Hirata, Travis Hudson, Robert Kleinmann, Neven Kresic, Philip LaMoreaux, David Lerner, John Moore, Nick Robins, Ramiro Rodriguez, Andrzej

Różkowski, Thomas Rude, Andrew Skinner, Jaroslav Vrba, Natalya Wiliams, Guido Wimmer, Alexander Zaporozec and Francois Zwahlen for their careful reviews and for their tremendous efforts in the linguistic correction of some manuscripts. We are very much obliged to Series Editor, Nick Robins, for his valuable suggestions and very kind help in editorial work. We would also like to express our special thanks to colleagues at the University of Silesia, particularly Dorota Grabala, Pitr Siwek and Jacek Wróbel for their valuable help with editing this publication.

Finally, we would like to express our special thanks to IAH and to UNESCO for their crucial financial support, without which the attendance at the conference as well as some contributions to this book by several experts from Eastern Europe and the Developing Countries, would not have been possible.

A.J. Witkowski, A. Kowalczyk & J. Vrba
March 2007

Foreword

A picture speaks more than a thousand words and a map more than thousand pictures. Maps, therefore become an important tool for scientists, managers, policy and decision makers and the public. The art of groundwater vulnerability mapping based on groundwater vulnerability assessment has developed historically from geological and hydrogeological mapping.

The term vulnerability of groundwater to contamination was introduced by French hydrogeologist J. Margat in 1968 and the first vulnerability map was constructed in France by M. Albined in 1970. Since the early 1980s more complex methods of groundwater vulnerability assessment have been developed and a considerable number vulnerability maps of various scales and objectives have been produced throughout the world.

The concept of groundwater vulnerability assessment is based on the assumption that

1. the physical environment may provide some degree of protection to groundwater against natural and human impacts, and
2. some land areas are more vulnerable than others.

Groundwater vulnerability portrayed on a map shows various homogeneous areas, as cells or polygons, which have different levels of vulnerability. The differentiation between the cells is, however, arbitrary because vulnerability maps only show relative vulnerability of certain areas to others, and do not represent absolute values.

Vulnerability of groundwater is a relative, non measurable, dimensionless property. Aggregating a number of key vulnerability attributes to one vulnerability class (index), involves the various steps of selection, scaling (transforming attributes into dimensionless measures), rating and weighting. A final groundwater vulnerability class is a mathematical aggregation of individual attributes across different measurement units so that the final vulnerability output is dimensionless.

A generally recognized and accepted definition of groundwater vulnerability has not yet been developed. However, there are no significant differences in the formulation of groundwater vulnerability between individual authors. Groundwater vulnerability is mainly formulated as "an intrinsic property (characteristics) of the groundwater (aquifer) system that depends on the sensitivity of that system to human and/or natural impacts", or "the sensitivity of the aquifer to being adversely affected by an imposed contaminant load", or "the intrinsic susceptibility of an aquifer to contamination".

There are generally two types of groundwater vulnerability assessments and maps: intrinsic and specific. Intrinsic vulnerability is based on the assessment of natural climatic, geological and hydrogeological attributes. Specific vulnerability relates to a specific contaminant, contaminant class, or human activity and is mostly assessed in terms of the risk of the groundwater system becoming exposed to contaminant loading. Recharge, soil properties, lithology and thickness of the unsaturated zone and depth to water table are the key attributes of both intrinsic and specific vulnerability. However, contaminants from point sources often enter the groundwater system beneath the soil profile (e.g. underground oil tanks, septic tanks) and the role of soil as an attenuation medium is then by-passed.

Some authors include the saturated aquifer in vulnerability assessment procedures, whilst other authors do not. However, the aquifer cannot be seen as a homogenous unit. Its vulnerability maybe significantly different at different places. The recharge area is always highly vulnerable and the discharge area, particularly in case of confined aquifers, is of low vulnerability. Aquifer hydraulic conductivity, does take a part in many groundwater vulnerability assessment procedures. Field and laboratory observations, however, demonstrate that the hydraulic conductivity of the aquifer may change significantly when fresh groundwater is replaced by polluted water.

The travel time of the contaminant should be included within specific vulnerability attributes. Some authors, however, state that the aquifer is equally vulnerable if contaminant travel time is one year or one hundred years and, therefore, the travel time attribute should be taken out of the vulnerability assessment concept.

The most important attribute in the assessment of specific groundwater vulnerability is the attenuation capacity of the physical environment (soil and rock) with respect to the properties of individual contaminants. However, the attenuation capacity will reduce with time, resulting in a changed vulnerability to the groundwater system. In the case of persistent and mobile contaminants, the contaminant travel time depends largely on the thickness and vertical permeability of the unsaturated zone. The amount of water stored in the aquifer and the net recharge, both control dilution of the contaminant in the aquifer.

Groundwater vulnerability maps are classified as problem oriented, specialized environmental maps derived from the basic hydrogeological map. Vulnerability maps are used for groundwater protection planning, management and decision making, for identification of areas susceptible to contamination and for public information and education. Basically two types of vulnerability maps exist. The intrinsic maps are used to evaluate the intrinsic groundwater vulnerability to a generic conservative pollutant. Single purpose and multipurpose maps are the main categories of specific vulnerability maps. In single purpose maps the vulnerability is evaluated with respect to only one type of contaminant or group of contaminants of similar properties. The multi-purposed maps are focused on presentation of various groups of contaminants of different properties which have been identified in the mapped area. The GIS format is widely used to present various vulnerability scenarios on the vulnerability maps.

The overall utility of a vulnerability map is dependent on the scale at which the map has been compiled. Selection of the optimal scale to aggregate and present the groundwater vulnerability information depends on the data availability and its reliability, the information needed by the user and the aim of the map. The maps are only as good as the information and data upon which they are based and the knowledge and experience of the map makers. It is important that disclaimers appear on maps, informing the user:

1. about the level of accuracy of the presented information.
2. of the map limitations and any restrictions on the intended use.

Vulnerability maps are living documents. Without periodical updating, the degree of potential map misuse and misinterpretation increases.

Many hydrogeologists agree on which groundwater vulnerability attributes are relevant, but they use different methodologies for combining these attributes into a vulnerability statement. Neither terminology nor approach is standardized. Given the same data base, different authors will not arrive at the same conclusions.

Groundwater systems are much too diverse to be subjected to a standardized vulnerability assessment. However, it seems important and topical to start the process of formalization:

1. Formulation of a consistent and widely accepted definition of groundwater vulnerability.
2. Development, formalization and implementation of methods and procedures of groundwater vulnerability assessment.
3. Unification of vulnerability symbols and legends and standardization of vulnerability classes on the maps, and to make vulnerability maps internationally comparable.
4. Definition of the general content of vulnerability maps – both intrinsic and specific.
5. Development or improvement of groundwater monitoring networks to acquire data for more precise vulnerability assessment and mapping.

The "Groundwater vulnerability assessment and mapping" conference and the topics discussed in that forum have provided significant support to the new developments in groundwater vulnerability assessment and mapping. Allow me to express my appreciation and thanks to the organizers and sponsors for the preparation and organization of the conference and to Andrzej Witkowski, president of the organizing committee, my admiration to his personal effort and enthusiasm to make this conference a reality.

J. Vrba
Chairman of IAH Commission
on Groundwater Protection

About the editors

Andrzej J. Witkowski (1950) received his MS (1973) and PhD (1983) from The University of Warsaw. He is Senior Lecturer at the University of Silesia, Poland, author and co-author of 90 scientific articles. He is certified professional hydrogeologist (American Institute of Hydrology and Polish Ministry of Environment), and consultant in many international projects. Since 2003 he serves as President of the International Mine Water Association (IMWA).

Andrzej Kowalczyk (1950) studied geology at the University of Warsaw (PhD degree at the Technical University of Wroclaw in 1983). He is presently Associate Professor and head of the Department of Hydrogeology and Engineering Geology at the University of Silesia, Poland. He participated in various projects in the field of hydrogeology and is author and co-author of 80 scientific articles.

Jaroslav Vrba (1932) received his MS and Doctorate from Charles University in Prague, Czech Republic. He has practiced hydrogeology in many countries for almost 40 years. At present he works in a private consulting practice and as UNESCO consultant for IHP. He chaired and coordinated many UNESCO-IHP groundwater projects and is author, co-author and editor of many articles and books. He is Past Vice President of IAH and Chairman of IAH Commission on Groundwater Protection; appointed honorary member of International Association of Hydrogeologists.

Keynote introductory paper

Aquifer pollution vulnerability concept and tools – use, benefits and constraints

Stephen Foster
IAH President

ABSTRACT: A brief overview of the development, use and limitations of the aquifer pollution vulnerability concept and related mapping tools is given. While such approaches are essentially pragmatic simplifications, if appropriately formulated they are capable of providing a scientifically consistent input to formalised groundwater pollution risk assessment, which is required for advancing implementation of the provisions of the EC-Water framework Directive and for raising public awareness of groundwater pollution hazards.

1   EVOLUTION OF POLLUTION VULNERABILITY CONCEPT

In an 'ideal world', the groundwater hazard from each potentially-polluting activity would be investigated individually – but this is not realistic, cost-effective nor adequate to communicate concerns in the vast majority of cases. The expression **aquifer pollution vulnerability** thus started to be used intuitively to convey concerns about pollution from the land surface in a general way from the early 1980s – initially in France (Albinet & Margat 1970).

From the late 1980s there were various attempts to formalize the definition of the expression and to develop related mapping systems (DRASTIC, GOD, SINTACS, etc.) (Aller et al. 1987, Foster 1987, Foster & Hirata 1988, Civita 1994). These all attempted to represent complex processes in a simple fashion, which was scientifically based, but involved different ranges of contributing factors, varying degrees of simplification and subjective professional judgement (Figure 1 illustrates the GOD aquifer pollution vulnerability system).

The critical question that faces those attempting such simplification is **the validity of using a single 'integrated vulnerability index'**. This bearing in mind that (in scientific reality) each class of potential groundwater contaminant will be influenced to different degree by various attenuation processes naturally operating in the soil and vadose zone (Foster & Hirata 1988). However, if we constrain use of the term 'vulnerability' to consider only potentially-polluting activities at the immediately-overlying land surface and use **'smart definitions' for vulnerability classes** (Table 1), this problem can be largely overcome and the use of an integrated index justified. This then greatly favours the practical application of the concept.

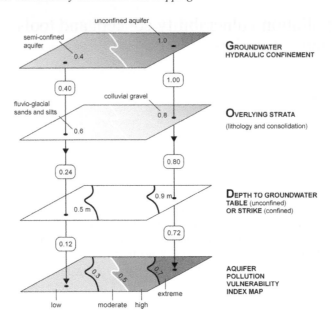

Figure 1.   Generation of aquifer pollution vulnerability maps using the GOD system.

Table 1.   Pragmatic definition of the relative classes of aquifer pollution vulnerability at any given location from activities on the immediately-overlying land surface.

Vulnerability class	Corresponding definition
Extreme	Vulnerable to most water pollutants with relatively rapid impact in many pollution scenarios
High	Vulnerable to many pollutants, except those strongly absorbed or readily transformed, in many pollution scenarios
Moderate	Vulnerable to some pollutants but only when continuously discharged or leached
Low	Only vulnerable to conservative pollutants in long-term when continuously and widely discharged or leached
Negligible	Confining beds present with no significant vertical groundwater flow (leakage)

2   CONTEXT FOR PRACTICAL APPLICATION OF MAPPING TOOLS

Although major simplification is involved, aquifer pollution vulnerability maps have become valuable tools in the following practical contexts:

- communicating concerns about the potential level of groundwater pollution hazard to civil society and the general public (in effect **making groundwater more visible**)
- providing a scientifically-based input to local land-use planning and effluent discharge control procedures
- as part of more **formalized groundwater pollution risk screening procedures**, especially where quality monitoring networks are still inadequate (e.g. for implementation of the EC–Water Framework Directive of 2000).

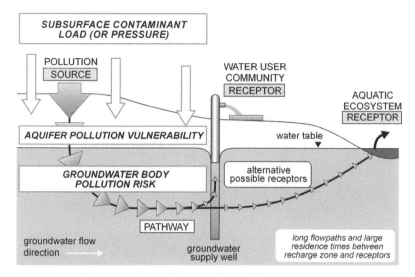

Figure 2.   General scheme of groundwater risk assessment for different receptors showing the preferred role and interactions of aquifer pollution vulnerability.

For all of the above applications the most important factor in selection or design of an aquifer pollution vulnerability mapping system will be **'fitness for purpose'** – such that the 'vulnerability zones' defined have strong resonance with environmental and land-use management decision-making.

In the latter of the above three applications, aquifer pollution vulnerability maps should preferably depict **only** the spatial variation of those intrinsic ground characteristics determining **potential vertical contaminant pathways** (Figure 2), since they have to be capable of interacting readily with:

- surveys of **subsurface contaminant loading** (or pressures upon the groundwater system).
- aquifer numerical modelling outputs delineating the area of groundwater capture or **potential horizontal pollutant pathway for various receptors** (Figure 2).

3   KEY FACTORS IN POLLUTION VULNERABILITY ASSESSMENT AND MAPPING

The central approach in all schemes of aquifer vulnerability assessment is to classify the (essentially intrinsic) characterisitics of the strata overlying the saturated aquifer (vadose zone or confining beds) according to their:

- physico-chemical characteristics to **retain and degrade potential water pollutants** and consequently contribute to contaminant attenuation capacity.
- physical properties which can **reduce the rate of vertical water infiltration**.

It is recognised that much emphasis needs to be placed on the likelihood of **preferential flowpaths developing in the vadose zone** (usually as a result of fracturing) (Foster 1997, Foster et al. 2002), in view of their significance in increasing aquifer pollution vulnerability

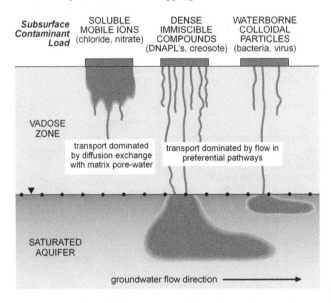

Figure 3.   Development of preferential flow in the vadose zone and its significance for aquifer pollution vulnerability.

to certain of the most serious classes of pollution (e.g.: from pathogens and immiscible organic compounds) (Figure 3).

Certain other questions arise when selecting or designing systems of aquifer pollution vulnerability assessment, especially those relating to the inclusion of factors to take account of:

- **pollutant attenuation in the saturated zone** – the author is of the opinion that this tends to add unnecessary complexity and confusion since it is not appropriate to assume a 'characteristic vertical travel distance to a typical receptor', and it is thus better to include consideration of this parameter in other ways
- **attenuation capacity in the soil profile** – bearing in mind that in reality this can be highly modified or completely removed (especially in urban areas), the author is of the opinion that a soil profile factor reducing vulnerability should only be included in rural areas and where it is clear that any pollutants will be leached from the soil and not 'injected' deeper into the sub-soil
- **natural infiltration rates to aquifer** – despite the fact that most contaminant transport depends upon water flux the author considers it counter-productive to include such a factor in aquifer vulnerability and better to consider all forms of recharge as an integral part of the subsurface contaminant load (or pressure), since in reality recharge is a variable parameter depend greatly on human activity, such as irrigation returns and urban drainage.

4   PRACTICAL LIMITATIONS OF POLLUTION VULNERABILITY MAPPING

It has to recognised that any scheme of aquifer pollution vulnerability assessment and mapping is subject to inevitable limitations relating to difficulty in adequately representing the complexity of field situations, particularly in the following instances (Foster et al. 2002):

- the **presence of semi-confined aquifer systems** – where considerable care will be needed to identify the aquifer horizon of interest for potable water-supply provision and assess the contaminant attenuation capacity of the overlying semi-confining beds bearing in mind that these may be geologically discontinuous
- the **situation along surface watercourses** – which will often have complex influent relationships to underlying aquifers and unknown streambed attenuation properties.

Moreover, aquifer pollution vulnerability mapping schemes do not normally take into account large-scale man-made physical disturbance of the vadose zone (as may occur in mining areas or through intensive urban infrastructure development) – such situations need to be identified at an early stage and mapped accordingly. Despite the major research advances in understanding sub-surface contaminant transport and attenuation mechanisms, there is still considerable uncertainty over the behaviour of some important types of pollutant (e.g. MTBE, PCBs, etc.) in certain classes of porous media (Foster 1998).

5 FROM AQUIFER POLLUTION VULNERABILITY TO GROUNDWATER POLLUTION RISK ASSESSMENT

To move from aquifer pollution vulnerability maps to groundwater pollution risk assessment, a systematic inventory of the interacting subsurface contaminant load (both existing and potential future) is essential (Foster & Hirata 1988). In EC circles the cumulative load is known as the **'contaminant pressure on the groundwater system'**. This is the often-neglected complementary component of groundwater pollution risk assessment, and standardised desk and field techniques are helpful for the required survey work (Foster et al. 2002). Considerable experience and insight will be required for a balanced survey and diffuse pollution sources usually present more difficulty to estimate than potential point sources.

Another key facet of groundwater pollution risk assessment is the **delineation of flow and capture zones around public groundwater supply sources** (and other groundwater receptors), which in effect provide an additional priority focus for the risk assessment process (Figure 4). The delineation of such zones (known as **perimeters** in some EU countries) is considered the best way of dealing with the capacity of aquifers for contaminant transport, dilution and attenuation in the saturated zone (Foster et al. 2002). They also provide the scientific basis for **potable groundwater quality protection**, and are required for the implementation of the EC Water Framework Directive of 2000. But they also have their limitations as a result of unstable geometries where:

- aquifer pumping regimes are hydraulically unstable
- karstic aquifers with swallow holes and caverns occur
- multi-layered semi-confined aquifers are present.

6 FROM GROUNDWATER POLLUTION RISK ASSESSMENT TO GROUNDWATER QUALITY PROTECTION

Groundwater pollution risk assessment provides a clearer appreciation of the actions needed to protect groundwater quality, and internationally should become an **essential component**

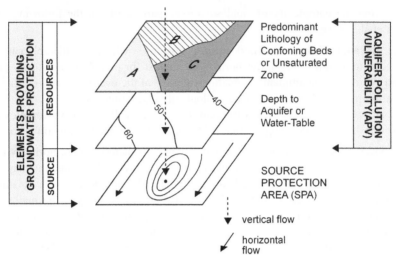

Figure 4.   Integration of aquifer pollution vulnerability mapping and delineation of source protection areas as the basis of a GIS system for groundwater quality protection.

of best practice in environmental management – the critical step being to 'make visible' the link between the use of given tracts of land and groundwater (resource and supply) quality (Foster et al. 2002). To protect aquifers against pollution it is necessary to constrain land-use and to control effluent discharge and waste disposal practices – and the process of land surface zoning will define a need to declare certain areas as especially critical for groundwater quality conservation and supply protection.

The legal feasibility of this approach will also have to be assessed. In **certain legal codes it is possible to control land-use activities in the interest of groundwater quality**, providing an overriding 'public interest' or 'strategic need' can be demonstrated – even without payment of compensation to affected land owners. But in some others powers will be more limited, and in some legal codes it will be more appropriate to **use groundwater zoning information as an input to the land-use planning process**, at least as far as controlling new sources of groundwater pollution is concerned.

Even where there is no quasi-legal basis for land-use planning to take account of groundwater interests, the procedures for **groundwater pollution hazard assessment constitute an effective vehicle for mobilising involvement of the relevant stakeholders**. The ultimate responsibility for groundwater pollution protection must lie with the relevant environment or water resource agency of national or local government, but given their responsibility to conform with codes of sound engineering practice, an obligation should also rest with **water-service companies to be proactive in promoting pollution hazard assessments** for all their groundwater sources (Foster et al. 2002), to provide a sound basis for forceful representations to be made for the implementation of necessary pollution control and aquifer protection measures.

7   CONCLUDING REMARKS

There is little doubt that the aquifer pollution vulnerability concept (and its practical manifestation in land surface mapping) is an extremely valuable tool for groundwater quality protection, when 'sensitively tuned' to the specific needs of a given application. Aquifer pollution vulnerability maps will play an increasingly important role as a key component of empirical screening methods for groundwater pollution risk assessment, which are needed to allow the implementation of the EC Water Framework & Groundwater Protection Directives especially in areas where actual monitoring of groundwater quality is limited and/or the potential time-lags of pollution incident impact in groundwater are very large.

ACKNOWLEDGEMENTS

The author would like to thank the IAH Polish National Group for the original suggestion to provide this overview, his co-worker Dr Ricardo Hirata of the Universidade de Sao Paulo – Brasil for sustained interest and work on these topics, and his assistant Gill Tyson for work on the production of the paper.

REFERENCES

Albinet, M. & Margat, J. 1970. Cartographie de la vulnerabilite a la pollution des nappes d'eau souterraine. *Bulletin BRGM 2nd* Series 3 (4), Orleans, France: 13–22.

Aller, L., Bennett, T., Lehr, J.H., Petty, R.J. & Hackett, G. 1987. DRASTIC: a standardized system for evaluating groundwater pollution potential using hydrogeologic settings. US-EPA Report 600/2-87-035, Washington DC, USA.

Civita, M. 1994. Le carte della vulnerabilita degli acquiferi all inquinamento: teoria e pratica. Pitagora Editrice. Bologna, Italy.

Foster, S.S.D. 1987. Fundamental concepts in aquifer vulnerability, pollution risk and protection strategy. *Proceedings International Conference VSGP*. Noordwijk, The Netherlands.

Foster, S.S.D. 1997. Groundwater recharge and pollution vulnerability of British aquifers – a critical overview. *Geological Society Special Publication* 13. London, UK: 7–22.

Foster, S.S.D & Hirata, R.A. 1988. Groundwater pollution risk assessment – a methodology using available data. WHO-PAHO-CEPIS Publication. Lima, Peru: p. 79.

Foster, S.S.D., Hirata, R.A., D'Elia, M. & Paris, M. 2002. Groundwater quality protection – a guide for water utilities, municipal authorities and environment agencies. World Bank Publication. Washington DC, USA: p. 101.

Examples of the European approach
to groundwater vulnerability assessment
and mapping

CHAPTER 1

The vulnerability paradox for hard fractured Lower Palaeozoic and Precambrian rocks

N.S. Robins, A.M. MacDonald & D.J. Allen
British Geological Survey, Wallingford, UK and Edinburgh, UK

ABSTRACT: The WFD criteria for chemical status require consideration of both the susceptibility of a groundwater body to pollution, and the susceptibility of the associated receptors to pollutants which have entered the groundwater body. Traditional concepts of aquifer vulnerability assess the groundwater body, but assessment of concentrations at specific points at the water table can better be undertaken by developing novel methodologies. One such approach is to remove recharge or aquifer productivity from the assessment and to focus only on transport and attenuation processes of pollution migration. When applied to fractured aquifer systems this reverses the vulnerability assessment from one of weakly permeable, low storage and, therefore, low vulnerability, to rapid transport, poor attenuation and high vulnerability. The new method provides assessment only of a single point at the water table but is only a part of the overall assessment process for a groundwater body. To help users of the maps, and avoid confusion, careful and clear definitions of vulnerability, including which receptors and pathways are addressed, are required at all times.

1   INTRODUCTION

One of the most significant pieces of European water legislation to be produced in recent years is the EU Water Framework Directive 2000/60/EC – "establishing a framework for Community action in the field of water policy" which was introduced in December 2000. The Water Framework Directive (WFD) expands the scope of protection to all waters (surface and groundwaters) with the aim of meeting specified environmental objectives, set out in WFD Article 4, by 2015.

In order to achieve the objectives and to manage surface and groundwater in an integrated way, the WFD introduces River Basin Districts (RBD) and requires that a River Basin Management Plan (RBMP) be produced for each District. The initial phase of the RBMP is the delineation of bodies of water within the RBD and their characterisation to assess their uses and the degree to which they are at risk of failing to meet the environmental objectives set for them. For groundwater bodies the Article 4 environmental objectives include the need to achieve good status. This has two components, quantitative and chemical status. The achievement of good chemical status involves meeting several criteria, including meeting quality standards and avoiding damage to receptors, of which associated surface waters and terrestrial ecosystems are specifically mentioned.

The risk assessment component of groundwater body characterisation therefore involves both an assessment of the nature and magnitude of pollution pressures and the likelihood that the pressure at the surface will adversely affect the underlying groundwater body to the extent that its chemical status in 2015 will be poor. This risk assessment will be affected by the groundwater body's vulnerability to pollution, and, given the WFD criteria for chemical status, two aspects of vulnerability need to be considered:

- the susceptibility of the groundwater body itself to pollution, and
- the susceptibility of the associated receptors to pollutants which have entered the groundwater body.

2   CONVENTIONAL AQUIFER VULNERABILITY

Aquifer vulnerability is a construct designed to help planners to protect aquifers as an economic resource. The concept of aquifer vulnerability combines the hydraulic inaccessibility of the saturated zone to the penetration of pollutants, with the attenuation capacity of the strata overlying the saturated zone as a result of physico-chemical retention or reaction of pollutants (Foster 1998). Attenuation includes biochemical degradation, sorption, filtration and precipitation. However, there has not yet been a general agreement on what the strict definition of vulnerability should be (Vrba and Zoporozec 1994), and the term vulnerability has come to mean different things in different contexts. The concept describes the likelihood of a general contaminant at or near the ground surface arriving at the water table or at a groundwater source.

Conventional methodologies involve indexing and weighting of relevant properties or GIS format overlays of relevant properties. The most popular indexing system is DRASTIC (Aller et al. 1987), which incorporates Depth to water [x5], natural Recharge rates [x4], Aquifer media [x3], Soil media [x2], Topographic aspect [x1], Impact effect of vadose zone [x5] and hydraulic Conductivity [x3] along with their respective weightings in brackets. Thus in a weakly permeable aquifer with relatively low recharge rates the vulnerability is low, whereas a more permeable aquifer with greater recharge potential which is exposed at surface is highly vulnerable and its groundwater is a significant resource. In this way Lower Palaeozoic and Precambrian shales, greywackes, crystalline and other fractured rocks with little if any matrix porosity have been classified as low vulnerability. The low vulnerability rating was not considered problematic since such aquifers contain little groundwater to be protected, and population densities and, therefore, pollution pressures in these areas were often low. The method has been successfully applied in a number of countries, but in each case shortcomings have been identified. Rosen (1994), for example, identified that DRASTIC tends to underestimate the vulnerability of fractured rocks.

An alternative indexing system in use in some countries is GOD, which considers Groundwater occurrence including recharge, Overall lithology and Depth to groundwater, also using a scoring system (Foster 1987).

The UK adopted a modified scheme based on overlaying geological characteristics of particular formations (Table 1) and the soil leaching potential to derive national aquifer vulnerability maps (Robins et al. 1994). These maps were strongly weighted towards aquifer permeability and also had a number of shortcomings. They only classified highly and moderately permeable aquifers into sub-classes based on three soil leaching classes and assumed that all

Table 1.   Geological classification of UK formations for aquifer vulnerability map series (after Robins et al. 1994).

Highly permeable aquifers	Moderately permeable aquifers		Weakly permeable rocks (aquifers)
Highly permeable formations usually with the known or probable presence of significant fracturing. Highly productive strata of regional importance.	Fractured or potentially fractured but with high intergranular permeability.	Variably porous/ permeable but without significant fracturing.	Formations with negligible permeability.
Includes: Chalk and Upper Greensand, Jurassic limestones, Jurassic sandstones, Magnesian Limestone, Lower Greensand, Carboniferous Limestone.	*Includes*: Coal Measures, Millstone Grit, Devonian sandstones and some igneous and metamorphic rocks.	*Includes*: River gravels, glacial sands and gravels, Palaeogene sands and gravels.	*Includes*: Clays, shales, marls and siltstones, Mercia Mudstone and most igneous and metamorphic rocks

weakly permeable rocks were not vulnerable as they had negligible hydraulic conductivity. Furthermore, the element of depth to the water table was excluded because inadequate data were then available, and any protective cover afforded by superficial tills or Clay-with-flint was ignored with the proviso that site-specific investigation of the superficial cover is needed to determine its specific pollutant transport characteristics. Nevertheless, these maps, which are available at 1:100,000 scale for England and Wales and parts of Scotland and at 1:250,000 for Northern Ireland, have become essential decision support tools for planning and other uses. The maps were aimed at protecting the most productive and heavily used aquifers in the UK. In this they have been successful in that they have been incorporated into the planning process, even though they are based on an incomplete set of parameters.

A judgement was made on the volcanic and igneous rocks of the Lower Palaeozoic and Precambrian as to the degree of permeability offered by them due to the presence of fractures. For the most part this was judged to be negligible, although some strata were considered to be moderately permeable. This decision was based on available understanding of these strata to sustain useable water supplies, but it dismissed most of these strata as weakly permeable and effectively non-water bearing. This judgement considered more than just the presence of fractures, but rather the presence of interconnected fractures that formed an aquifer unit with at least some storage and in which groundwater flow and pollutant transport could take place. It also recognised that glaciation had removed any higher permeability regolithic material except in the periglacial south of England, e.g. the Cornish granites.

3   GROUNDWATER VULNERABILITY UNDER THE WFD

The WFD upsets the conventional aquifer vulnerability philosophy by designating as a groundwater body (the basic groundwater management unit) any geological formation capable of supplying at least $10\,m^3\,d^{-1}$ of water. This is a very low threshold and even weakly permeable rock units contain groundwater that needs to be protected, and a new approach is required to assign its vulnerability class, in order to facilitate the risk assessment component of groundwater body characterisation.

In addition, it must be borne in mind that the WFD characterisation risk assessment involves both the susceptibility of the groundwater body and its associated receptors to pollution. In this case receptors may be a wetland or flowing surface watercourse, both of which may also receive an additional pollutant load from surface runoff. However, surface runoff can be disregarded in the present context which concerns only the pollutant load arriving at the receptor in the form of groundwater baseflow. The WFD also requires the degree of pollution within a groundwater or surface water body to be assessed as a concentration that can be compared with a quality standard (WFD Annex V 2.3.2).

Novel approaches to vulnerability are likely to be required to address the requirements of the Water Framework Directive. New approaches are being made, some based on disassembling conventional vulnerability assessment. This enables bespoke vulnerability assessments to better target the precise needs of the WFD as stated in Article 5, characterisation. One such approach is to remove the element of recharge, or conversely productivity, from the vulnerability assessment and concentrate solely on the transport and attenuation aspects of the pollutant-pathway-receptor concept. The importance of recharge in the vulnerability equation is thus downgraded, as the resource potential or economic value of the groundwater is no longer deemed to be a significant factor. This argument hinges on the WFD requirement to consider concentrations of pollutants against a set of quality standards. It recognises that some groundwaters may have intrinsic value, despite their lack of economic significance.

This approach, therefore, recognises the vulnerability of groundwater in, for example, a single fracture in an exposed basement aquifer. Whether the whole groundwater body, considered as a receptor, is vulnerable (i.e. whether concentrations of pollutant are likely to be elevated across the entire body) will then depend on issues such as the degree of interconnection of the fractures and whether pollution is diffuse or point-source in origin. But in order to consider the groundwater body vulnerability with respect to associated receptors such as rivers receiving baseflow, the recharge component again becomes relevant.

The paradox arises when dealing with the locally important aquifers of Lower Palaeozoic and basement rocks, which generally have small quantities of groundwater circulating in shallow fractures that may or may not be interconnected. Compounded by a thin cover of permeable soil the vulnerability assessment of these rocks, once recharge is removed from the formulae, is now reversed in the new assessment from not vulnerable to highly vulnerable. The fractured rocks are now perceived as highly vulnerable with an easy and rapid pathway to the saturated zone with little opportunity for attenuation. This is because each fracture may be vulnerable to pollution at a single point on the ground surface above it, although neither the overall groundwater body, nor baseflow discharging from it, need be overly contaminated, i.e. some of the groundwater and, therefore, parts of the groundwater body, are vulnerable but the overall aquifer is not.

In such weakly permeable aquifers the volume of recharge and the available productivity are extremely small and the relative volumes of polluted groundwater baseflow are small and may be insignificant. The paradox is most apparent when this output is compared to a conventional (with recharge) vulnerability assessment such as the DRASTIC type maps which clearly indicate that weakly permeable fractured rocks are not vulnerable to pollution. This has the potential to confuse users and planners. Although it is laudable to disassemble the vulnerability assessment process in order to attain procedural transparency, it is dangerous to leave it incomplete unless it is properly labelled, as this can mislead and confuse. Nevertheless the disassembling process is the logical way to deal with the WFD. Although groundwater vulnerability is important to know, conventional aquifer vulnerability is generally the more useful concept for land use planning – we must bear in mind the planners who use the final maps.

4   FRACTURE VARIABILITY

There is a range of fractured rocks that occur in Palaeozoic and older rocks and in volcanic and igneous rocks of any age. Tellam & Lloyd (1981) demonstrated the variability in hydraulic properties that occur in such rocks and reported field hydraulic conductivities for Palaeozoic mudrocks as low as $10^{-6} \, \mathrm{m \, d^{-1}}$, but as high as $10^{-1} \, \mathrm{m \, d^{-1}}$. Robins and Misstear (2000) divided fractured basement rocks into two distinct categories (Table 2) each with discreet hydrogeological characteristics. Any scheme of vulnerability assessment that attempts to lump all secondary permeable rocks into one class will encourage an overstatement of vulnerability. However, the heterogeneous nature of these strata is also significant, even for the regionally important Carboniferous Limestone. This is illustrated by borehole performance data for 225 boreholes in the west of Ireland which illustrate a large range in yield which is skewed towards the lowest values (Table 3).

The variety of fractured rocks coupled with a range of unsaturated depths indicate that a range in vulnerability is possible. Many workers allude to this, Morris et al. (2003), for example, state that extreme vulnerabilities are associated with highly fractured aquifers with a shallow water table as they offer little chance for contaminant attenuation.

Table 2.   Principal aquifer characteristics in upland hard rock areas.

Class	Geology	Properties	Transmissivity $(m^2 \, d^{-1})$	Storativity	Borehole yield $(l \, s^{-1})$
Regionally important aquifers	Carboniferous Limestone, Devonian, some volcanic rocks	Anisotropic; fracture flow dominant; regional and local flow paths	100–4000	0.01–0.20	5–40
Locally important aquifers	Precambrian and Lower Palaeozoic; some Upper Palaeozoic; some volcanics	Anisotropic, secondary porosity dominant, local flow paths	20–50	<0.05	1–5

Table 3.   Summary of performance characteristics of 225 boreholes in Carboniferous Limestone in the west of Ireland (after Drew and Daly, 1993).

Variable	Maximum	Minimum	Mean
Depth (m)	177	3	57
Yield $(l \, s^{-1})$	76	0	2.4
Specific capacity $(l \, s^{-1} \, m^{-1})$	7.6	0	0.8

The relationship between attenuation and fracture flow must be considered carefully. It is tempting to dismiss attenuation in vadose fracture systems because of the perceived speed of transport from surface to water table. In many situations attenuation can take place in the soil horizon and continue to some extent in poorly dilated factures below. Even thin soils may contain some organic carbon, and this provides an active zone for ion exchange and sorption to take place. Peat is a particularly attractive medium for chemical activity, and this may continue in the underlying fracture system. Attenuation is, therefore, possible in some fracture systems. However, the soil can easily be bypassed (e.g. from contaminant sources such as septic tanks etc.) or the soil may be absent. The precautionary principle for fractured rocks, however, suggests that all fracture systems be considered vulnerable until they can be demonstrated otherwise.

5   CONCLUSION

The concept of groundwater vulnerability is evolving. The EU Water Framework Directive demands that all groundwater is protected and states that pollutant concentrations in groundwater should be used in the risk assessment for groundwater bodies. This consideration has led to the deconstruction of the groundwater vulnerability concept in order to focus on pollution concentrations arriving at the water table. However, this methodology is appropriate to only one part of the WFD pollution risk assessment, which requires both the pollutant risks to the whole groundwater body and to other associated receptors to be evaluated.

The new methodology establishes the potential for groundwater at the water table directly beneath a point on the ground surface to have elevated concentrations of contaminants. For example, groundwater within fractures in granite may be highly vulnerable to contamination from a soakaway located directly above it. The original concept of aquifer vulnerability, however, includes factors (e.g. recharge and aquifer permeability) which are appropriate for the assessment of the susceptibility of associated receptors – such as streams – to groundwater pollution and the method has proven to be valid as a planning tool to protect important aquifers from contamination.

Both vulnerability methodologies, therefore, have their place. However, in order to avoid confusion, groundwater vulnerability methodologies specifically designed as tools to assist in the characterisation of groundwater bodies should be distinguished from existing published vulnerability maps. For both, careful and clear definitions are required concerning which receptors and pathways are addressed. In the meantime it may be better to refer to the new methods developed for the WFD as *Groundwater vulnerability screening tools for the Water Framework Directive* rather than aquifer vulnerability maps.

ACKNOWLEDGEMENT

The authors are grateful for valuable discussions with colleagues John Chilton and Brian Adams. This paper is published by permission of the Director British Geological Survey (NERC).

REFERENCES

Aller, L., Bennett, T., Lehr, J.H., Petty, R.J. & Hackett, G. 1987. DRASTIC: a standardised system for evaluating groundwater pollution potential using hydrogeologic settings. US-EPA Report 600/2-87-035.

Drew, D. & Daly, D. 1993. Groundwater and karstification on mid-Galway, south Mayo and north Clare. Geological Survey of Ireland Report RS 93/3, Dublin.

Foster, S.S.D. 1987. Fundamental concepts in aquifer vulnerability, pollution risk and protection strategy. In: van Duijvenbooden W & van Waegeningh HG (eds.) *Vulnerability of soils and groundwater to pollution*. TNO Committee on Hydrological Research, The Hague, Proceedings and Information, 38: 69–86.

Foster, S.S.D. 1998. Groundwater recharge and pollution vulnerability of British aquifers: a critical review. In: Robins NS (ed.) *Groundwater Pollution, Aquifer Recharge and Vulnerability*. Geological Society, London, Special Publications, 130: 7–22.

Morris, B.L., Lawrence, A.R., Chilton, P.J., Adams, B., Calow, R.C. & Klinck, B.A. 2003. Groundwater and its susceptibility to degradation, a global assessment of the problem and options for management. United Nations Environment Programme, Nairobi.

Robins, N.S, Adams, B., Foster, S.S.D. & Palmer, R. 1994. Groundwater vulnerability mapping: the British perspective. *Hydrogèologie*, 3: 35–42.

Robins, N.S. & Misstear, B.D.R., 2000. Groundwater in the Celtic regions. In: Robins N.S. & Misstear B.D.R. (eds.) *Groundwater in the Celtic regions: a study in hard rock and quaternary hydrogeology.* Geological Society, London, Special Publications, 182: 5–17.

Rosen, L. 1994. A study of the DRASTIC methodology with emphasis on Swedish conditions. *Ground Water*, 32: 278–286.

Tellam, J.H. & Lloyd, J.W. 1981. A review of the hydrogeology of British onshore non-carbonate mudrocks. *Quarterly Journal of Engineering Geology*, 14: 347–355.

Vrba, J. & Zoporozec, A. (eds.) 1994. Guideline on mapping groundwater vulnerability. International Association of Hydrogeologists, Hannover, International Contributions to Hydrogeology, 16.

CHAPTER 2

Evaluation of reactive transport parameters to assess specific vulnerability in karst systems

M. Sinreich, F. Cornaton & F. Zwahlen

Centre of Hydrogeology (CHYN), Neuchâtel University, Switzerland

ABSTRACT: The VULK code is an analytical one-dimensional transport solver that has been developed for quantitative vulnerability assessment in karst systems. For this purpose, it has been provided with a dual-porosity approach for accounting for preferential flow in enlarged fissures and karst conduits. Specific vulnerability evaluation requires additional information on retardation and degradation processes in the subsurface besides flow characteristics. An improved VULK version has the capacity to simulate an array of physical and geochemical reactions, which may affect specific contaminants. Adsorption, cation exchange or precipitation may be modelled with VULK, either by using linear equilibrium or kinetic sorption approaches. Processes such as biodegradation or die-off causing the elimination of contaminant mass may use a first-order decay term. VULK's multi-process approach thus allows fate and transport modelling of many kinds of contaminant while classifying specific vulnerability and risk in terms of key physical parameters, such as transit time, attenuation (relative concentration) or recovery. Calibration and validation of contaminant-specific transport simulations can be achieved using comparative tracer testing.

1 INTRODUCTION

According to the European COST Action 620 definition dealing with *Vulnerability and risk mapping for the protection of carbonate (karst) aquifers* (Zwahlen ed. 2004), intrinsic vulnerability of groundwater to contaminants accounts for the geological, hydrological and hydrogeological characteristics of an area, but is independent of the nature of the contaminants and the contamination scenario. Specific vulnerability takes into account the properties of a particular contaminant or group of contaminants in addition to an area's intrinsic vulnerability when considering threats posed to water quality by specific contaminants. Risk may be defined as the probability of harmful groundwater contamination with respect to a combination of (intrinsic or specific) vulnerability, hazard potential and adverse consequences on groundwater's social and economic value.

Following the origin-pathway-target model (DoELG/EPA/GSI 1999 and Daly et al. 2002), contaminants released at the land surface migrate through the unsaturated zone layers until they reach the water table (resource protection). However, in karst systems contaminants can be preferentially transferred along enlarged fissures and solution conduits without interaction with aquifer surfaces, thus bypassing the protective cover. If vulnerability assessment aims to protect a spring or a pumping well (source protection), the lateral

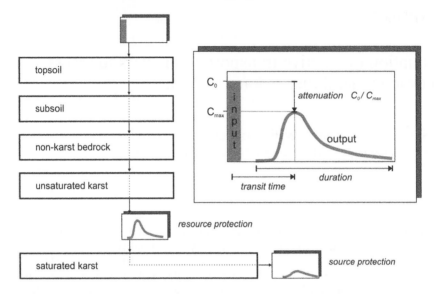

Figure 1.   Multi-layer use of the VULK model for resource and source protection purposes. Dominant transit time and attenuation (inverse of relative maximum concentration) are used as key output parameters.

pathway followed by groundwater through the saturated karst also needs to be incorporated (Figure 1).

Brouyère et al. (2001) proposed a quantitative definition of groundwater vulnerability using a physically-based approach. Using this approach, advective-dispersive transport modelling can be considered to be an adequate tool for carrying out intrinsic vulnerability assessment (Jeannin et al. 2001). This provides key flow parameters, such as transit time from infiltration point to the target, attenuation of a unit input concentration, or the duration of breakthrough at the target (Figure 1) Pochon et al. (2004) implemented this theoretical approach of the field scale by using the VULK model. Based on these achievements, this quantitative approach has now been incorporated into contaminant-specific vulnerability and risk assessment using a reactive transport version of the VULK groundwater modelling tool.

2   MODEL DESCRIPTION

The VULK code, developed at the CHYN, is an analytical one-dimensional transport solver (steady-state flow, transient transport) that has been created specially for intrinsic vulnerability assessment in karst settings (Jeannin et al. 2001, Cornaton et al. 2004). The model has the capacity to generate breakthrough curves resulting from a hypothetical unit contaminant input passing through a defined multi-layer system (Figure 1). Analytical solutions are derived by applying a Laplace transform that can perform successive convolutions of each output signal routed through a given layer using the transfer function for the next layer. Final results are obtained using a numerical inversion algorithm. For a more detailed mathematical description of this process the reader is referred to Jeannin et al. (2001).

VULK has now been upgraded to allow it to be used for specific vulnerability assessment by employing analytical solutions of the 1D advection-dispersion-reaction equation (1). The additional retardation and degradation terms allow contaminant-specific processes to be taken into account:

$$\frac{\partial C}{\partial t} = D \cdot \frac{\partial^2 C}{\partial x^2} - v \cdot \frac{\partial C}{\partial x} - \frac{\rho}{\Phi} \cdot \frac{\partial C_{sorb}}{\partial t} + \left(\frac{\partial C}{\partial t}\right)_{deg} \tag{1}$$

$$\text{dispersion} \quad \text{advection} \quad \text{retardation} \quad \text{degradation}$$

with
$$\frac{\partial C_{sorb}}{\partial t} = K \cdot \frac{\partial C}{\partial t} \tag{2}$$

for linear equilibrium sorption, where K is the distribution coefficient [L^3/M], and

$$\frac{\partial C_{sorb}}{\partial t} = k \cdot (K \cdot C - C_{sorb}) \tag{3}$$

for linear rate-limited (kinetic) sorption, where k is a first-order sorption rate constant [1/T] and the expression in brackets reflects deviation from equilibrium (e.g. Jury & Roth 1990). The terms ρ and Φ in equation (1) are bulk density [M/L^3] and porosity [$-$], respectively, D is the dispersion coefficient [L^2/T], and v is the average linear groundwater velocity [L/T].

Degradation is assumed to follow a first-order decay/transformation relationship using the decay constant λ [1/T]:

$$\left(\frac{\partial C}{\partial t}\right)_{deg} = -\lambda \cdot C \tag{4}$$

To make the model consistent with hydraulic characteristics particular to karst settings, VULK uses a dual-porosity solution for first-order mass transfer (Figure 2). Water flow through karst often occurs as preferential flow in conductive drains, whereby thick unsaturated zones may be rapidly by-passed (mobile phase). In such cases, only a second porosity type with significant water storage characteristics can enable specific attenuation processes to occur. The fundamental assumption made using the dual-porosity approach is that flow in the second porosity can be neglected (immobile phase). Retardation and degradation modules can be separately applied to both mobile and immobile phase domains. Hence, VULK simultaneously handles physical and sorption non-equilibria. Neville et al. (2000) provide a comprehensive overview of such model conceptualisation. Both retardation and dual porosity are suited to simulating tailing effects.

The key assumptions made for the retardation and degradation parts of the model are:

- Layer properties are uniform and time-independent
- Equilibrium sorption is instantaneous, reversible and governed by a linear sorption isotherm
- Rate-limited sorption is represented by a first-order reaction model
- Degradation is modelled as a first-order decay process and is identical for sorption sites and in the solute.

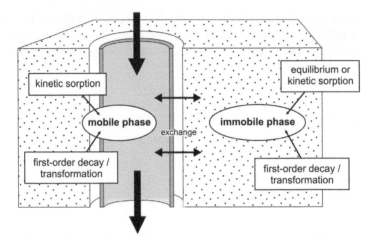

Figure 2.   Conceptual model of the dual-porosity approach with separate process simulation for each domain.

Table 1.   VULK input parameters (index *m* for mobile domain, index *im* for immobile domain).

	Reactive transport			
	Conservative transport			
single porosity	Flow velocity	v	Bulk density	ρ
	Layer thickness	d	Equilibrium sorption coefficient	K_m
	Dispersivity	α_L	Sorption rate constant	k_m
	Dilution factor	dil	Fraction of equilibrium sorption sites	F_m
			Decay constant	λ_m
dual porosity	First porosity	Φ_m	Fraction of mobile sorption sites	f
	Second porosity	Φ_{im}	Equilibrium sorption coefficient	K_{im}
	Exchange coefficient	α	Sorption rate constant	k_{im}
			Fraction of equilibrium sorption sites	F_{im}
			Decay constant	λ_{im}

For conservative transport, VULK requires each layer to have four input parameters for single porosity simulation, and seven for dual porosity simulation (Table 1, first column). Up to 10 additional variables may be applied when dealing with reactive contaminants (Table 1, second column).

3   MODEL APPLICATION

3.1   *Transport simulation*

VULK can be applied to simulate the fate and transport of specific contaminants according to the processes that affect the behaviour of these substances. Different kinds of physical and geochemical processes can be assigned using fundamental retardation and degradation equations, so that the model can become very flexible in terms of contaminant attenuation processes.

3.1.1   *Retardation*

Reversible sorption processes, such as adsorption, cation exchange or precipitation may be modelled either by using an equilibrium or a kinetic approach. A kinetic approach is indicated if residence time is significantly lower than the time needed for reactions to reach steady state. Only if processes occur instantaneously or if a sufficient time span is available, can equilibrium conditions be considered. Equilibrium sorption coefficients (K_m, K_{im} in Table 1) derive from a linear isotherm assumption and are prevalent in the literature for many geological settings and contaminant types. Kinetic sorption modelling is, in addition, governed by sorption rate constants (k_m, k_{im}). For each domain, a fraction of instantaneous sorption sites can be inserted (F_m, F_{im}). However, to separate the proportion of instantaneous and rate-limited sorption sites is believed to be excessively detailed for the purpose of vulnerability assessment. It is assumed that many sorption processes in the matrix (immobile phase) are subjected to equilibrium sorption conditions ($F_{im} = 1$), whereas a time-dependent sorption model is often indicated for the flow in conductive (macro)pores and conduits (mobile phase, $F_m = 0$). Nevertheless, the portion of sorption sites f comprising the mobile phase must always be evaluated when using the dual-porosity approach ($f = 1$ for single porosity).

3.1.2   *Degradation*

While the intrinsic vulnerability assessment approach assumes conservation of mass, specific processes may cause mass loss (recovery < 1). Processes resulting in contaminant elimination are modelled by employing first-order decay/transformation terms; processes include biodegradation, oxidation, reduction, decay or die-off. All these reactions are modelled with time-dependent concentration decreases. The VULK model does not distinguish between mass loss rate from the liquid phase and the sorbed phase; this may not always be consistent with natural reactions, e.g. when dealing with microbial inactivation. However, VULK is capable of differentiating between degradation in the mobile and in the immobile phase by applying decay constants for both domains. An analogous first-order decay function can be used to simulate irreversible kinetic sorption, which by definition does not cause retardation. Rate constants may then be defined in accordance to the sorption characteristics.

3.1.3   *Model input/output*

A variety of parameter combinations for significant attenuation processes is possible using the multi-process approach. A maximum of 0 reactive transport parameters, as listed in Table 1, may be used in model calibration, although not all of these are normally available for field applications. As a result of the above simplifications, VULK may be employed with a restricted number of input parameters. The modules provided for reactive transport may be further reduced in accordance to the properties of a selected contaminant. Neglecting both retardation and degradation terms finally allows a non-reactive (conservative) contaminant representing intrinsic vulnerability to be modelled.

The model visualises breakthrough curves for each sub-system (different layers; mobile/immobile phases) and gives key output parameters, such as transit time, maximum relative concentration, duration and recovery. Figure 3 illustrates the effect of specific processes on breakthrough shapes. The uppermost soil layer often plays a pronounced role

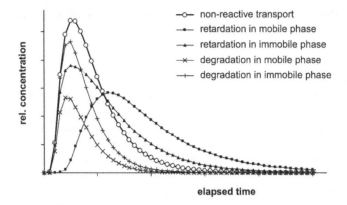

Figure 3.   VULK transport simulation through one layer with retardation and degradation effects.

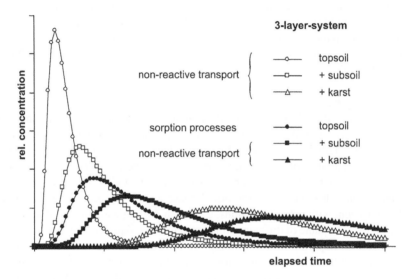

Figure 4.   Propagating effect of sorption in the uppermost layer compared to entirely non-reactive multi-layer transport.

in contaminant attenuation, an effect that is further propagated through the simulated multi-layer system (Figure 4).

3.2   *Vulnerability and risk mapping*

Since the main objective of the VULK code is to assess intrinsic and specific vulnerability, the latest version of the model has been provided with a direct link to a GIS database. Modelled breakthrough curves can be created for any point or polygon in a catchment that has been characterised in terms of input parameters. The curves provide values for key physical parameters, which can be displayed on a map, such as that completed by Pochon et al. (2004) for non-reactive transport. The results can assist in the compilation of resource protection or source protection maps. It should be noted that, despite the physically-based

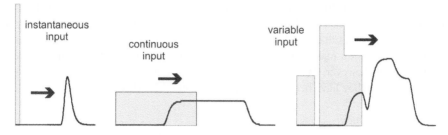

Figure 5.   Alterable VULK boundary conditions and related breakthrough curves.

character of the assessment, it is nevertheless no more than a first screening and not a detailed study. Nonetheless, input parameter evaluation needs to be accurate and representative, especially in karst areas where the heterogeneity of such settings must be taken into account. Note that VULK only considers the function of the protective layers, whereas a comprehensive vulnerability assessment in karst environments also requires rainfall and runoff phenomena to be incorporated (Daly et al. 2002).

With respect to the requirements of a progressive landuse management strategy, tools for contaminant-specific risk assessment are preferable. Risk evaluation includes consideration of hazard distribution and contaminant impact intensity, as well as the consequences of a contamination event. VULK can be used to simulate different contamination scenarios by considering not only the contaminant type, but also the contaminant quantity and the temporal release pattern. The user can define boundary conditions in terms of input concentration and duration (Figure 5). In this case, VULK provides absolute output concentrations with respect to the input stress, instead of relative concentrations, as is the case for a unit input signal in vulnerability assessment.

Impact maps, displaying landuse in terms of quantitative contaminant release, may thus serve as model input information. Groundwater quality thresholds, which indirectly incorporate contaminant toxicity, can be used as indicators for social and economic consequences of a groundwater contamination event.

3.3   *Vulnerability and risk classification*

Key physical parameters, including transit time and concentration attenuation, can be used to describe reactive transport in an analogous manner to conservative transport (Pochon et al. 2004). This is particularly important in karst systems, where transit time alone may not be sufficient to differentiate between particular contaminant types.

3.3.1   *Vulnerability assessment*

All other things being equal, contaminants released on a less vulnerable surface will have a longer transit time and be more attenuated than those released on a more vulnerable surface. Aquifers will be less vulnerable to sorbing and degradable contaminants than to conservative substances. Measured parameters can be used to generate intrinsic and specific vulnerability classes. Special parameter combinations and class limits may be indicated for particular assessment purposes. Figure 6 shows one possible vulnerability classification scheme using transit time ranges (3 days, 30 days, ~1 year, ~10 years) and attenuation ranges (10, 100, 1,000, 10,000 times) on a logarithmic scale as proposed by COST Action 620 (Zwahlen ed. 2004). This classification scheme assumes that a gradual increase in the physical parameters results in a decline in vulnerability. Vulnerability maps may be more meaningful to decision makers and other non-specialists than transit time maps or attenuation maps. Note, however, that parameters used in the assessment process lose their physical significance once they have been transferred to vulnerability classes.

3.3.2 Risk assessment

Risk assessment incorporates the consequences of groundwater contamination by considering chemical compounds' toxicities. The diagram displayed in Figure 7 represents an

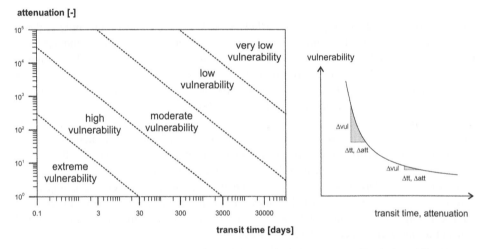

Figure 6.   Multiplicative combination of transit time and attenuation to define vulnerability classes (left). The classification is more sensitive towards high vulnerability values (right).

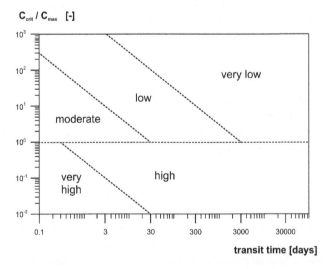

Figure 7.   Example of risk classification for chemicals with contaminant concentration as the crucial parameter (C_{crit}/C_{max}: inverse of maximum output concentration relative to threshold value).

example for a risk classification. The example illustrates how, when a critical concentration value is considered, category boundaries are offset compared to the vulnerability classes. The risk diagram could be used to decide that the risk associated with a contamination incident is regarded as high once the maximum output concentration is predicted to exceed the associated groundwater quality threshold.

4 MODEL CALIBRATION AND VALIDATION

Several data are needed to execute contaminant-specific VULK simulations, although many input parameters may be omitted. Model input data can be obtained from different sources, such as

- Geological and morphological mapping, geophysical surveys
- Artificial tracer experiments
- Spring monitoring, hydraulic tests
- Soil analysis, batch and column tests
- Literature review.

Tracer testing yields real breakthrough curves, which can provide control data that may be used for model calibration under realistic conditions. Tracer test results can be compared directly with synthesised breakthrough curves. Simultaneous injection of different tracers, to simulate the transport of particular contaminants, allows for comparative model calibration as well as validation of specific vulnerability maps generated using VULK at particular point. However, tracer experiments always reflect the overall role of the system, which can include several layers (Figure 8). Tracer testing involving a reduced number of passed layers lessens the ambiguity associated with experimental results. Laboratory measurements provide a more promising means of identifying retardation and degradation parameters (e.g. distribution coefficient, biodegradation rate). These data may thus be more confidently used for model parameter definition than inversely-modelled parameters, although up-scaling problems may arise.

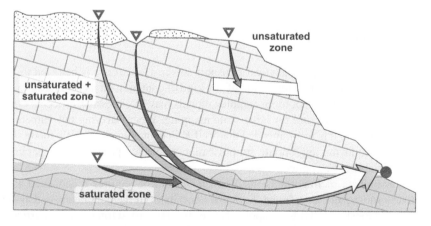

Figure 8.   Different kinds of possible field tracer experiments comprising one or several layers for VULK model calibration and validation.

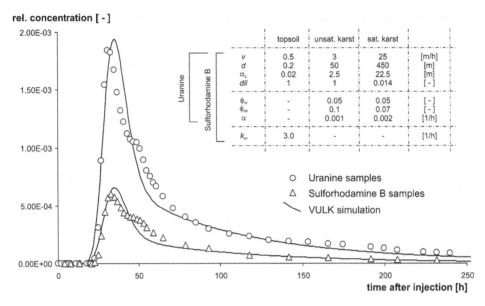

Figure 9.   Results of comparative tracer testing and modelled VULK breakthrough curves for con-servative transport (Uranine) and reactive transport (Sulforhodamine B).

Figure 9 represents the results of a comparative tracer test performed in the Swiss Jura Mountains (Pochon et al. 2004). Uranine and Sulforhodamine B were applied simultan-eously onto a karrenfield covered with a thin layer of soil, and were shortly thereafter detected at a nearby karst spring. The breakthroughs of both dyes were modelled using VULK assum-ing a 3-layer-system and ignoring long-term water storage in the epikarst zone. Uranine breakthrough was considered to represent a conservative substance, while Sulforhodamine B was regarded as a surrogate for a sorbing contaminant (Flury & Wai 2003). By assum-ing the same flow parameters for both substances, the curve simulated using VULK was fitted to the reactive tracer breakthrough curve by estimating an irreversible sorption rate for the thin topsoil layer.

5   CONCLUSIONS

VULK is a relatively simple analytical transport solver. However, it is believed to be an appropriate tool for modelling migration of any kind of contaminant using a multi-process approach. The requirements of more complex models may be excessive, especially in het-erogeneous environments and for large-scale mapping. When carrying out vulnerability and risk assessments, it is recommended that VULK in contaminant-specific mode should be used only employing appropriate simplifications. Selected processes and parameters must be justifiable to be applied.

Each contaminant of importance for basin management requires a dedicated assess-ment. Every subsequent assessment, however, benefits from the parameters and assump-tions once established. VULK is applicable to any type of setting, but is particularly convenient for karst systems due to the dual-porosity approach provided. The steady-state saturated flow assumption adequately reflects concentrated recharge conditions typical of karst aquifers. Although input data have sometimes to be roughly estimated, the advantage of vulnerability assessment using VULK is the calibration traceability and the physical logic behind the quantitative approach. Voigt et al. (2003) emphasised that quantitative methods should be further promoted in order to make vulnerability assessment less arbitrary. Physically-based methods thus provide meaningful data for the validation of qualitative methods based on empirical approaches.

Nevertheless, model assumptions are frequently unrealistic. In the strict sense, vulnerability and risk maps are only valid for the hydrologic and physical-chemical conditions assumed. For instance, processes assumed to be irreversible may reverse under changing conditions. Since these effects normally happen after the main contaminant breakthrough, they are not relevant to vulnerability assessment in karst environments.

Further studies using VULK for specific vulnerability assessment require the model to be based on field and laboratory measurements if better calibration is to be achieved. An application to a Swiss karst basin is foreseen as a means of creating vulnerability and risk maps for different contaminants of agricultural and industrial origin. Investigation of contaminant behaviour in karst settings and simulation of contamination scenarios by means of tracer experiments are currently ongoing.

REFERENCES

Brouyère, S., Jeannin, P.-Y., Dassargues, A., Goldscheider, N., Popescu, I.C., Sauter, M., Vadillo, I. & Zwahlen, F. 2001. Evaluation and validation of vulnerability concepts using a physical based approach. 7th Conference on Limestone Hydrology and Fissured Media, Besançon, *Sci. Tech. Envir. Mém. H. S.*, 13: 67–72.

Cornaton, F., Goldscheider, N., Jeannin, P.-Y., Perrochet, P., Pochon, A., Sinreich, M. & Zwahlen, F. 2004. The VULK analytical transport model and mapping method. In:*Vulnerability and risk mapping for the protection of carbonate (karst) aquifers* (Zwahlen ed.). Final report (COST Action 620).

Daly, D., Dassargues, A., Drew, D., Dunne, S., Goldscheider, N., Neale, S., Popescu, I.C. & Zwahlen, F. 2002. Main concepts of the "European approach" to karst-groundwater-vulnerability assessment and mapping. *Hydrogeol. J.*, 10: 340–345.

DoELG/EPA/GSI 1999. Groundwater protection schemes. – Department of the Environment and Local Government, Environmental Protection Agency, Geological Survey of Ireland: p. 24.

Flury, M & Wai, N.N. 2003. Dyes as tracers for vadose zone hydrology. *Reviews of Geophysics*, 41/1: 1–37.

Jeannin, P.-Y., Cornaton, F., Zwahlen, F & Perrochet, P. 2001. VULK: a tool for intrinsic vulnerability assessment and validation. 7th Conference on Limestone Hydrology and Fissured Media, Besançon, *Sci. Tech. Envir. Mém. H. S.* n° 13: 185–190.

Jury, A.J. & Roth, K. 1990. Transfer functions and solute movement through soil. Theory and applications. Basel: Birkhäuser.

Neville, C.J., Ibaraki, M. & Sudicky, E.A. 2000. Solute transport with multiprocess nonequilibrium: a semi-analytical solution approach. *J. Cont. Hydrol.*, 44: 141–159.

Pochon, A., Sinreich, M., Digout, M. & Zwahlen, F. 2004. Vaulion test site, Jura Mountains, Switzerland; Intrinsic and specific vulnerability mapping. – In: Zwahlen, F. (ed.) *Vulnerability and risk mapping for the protection of carbonate (karst) aquifers*; Final report (COST Action 620).

Voigt, H.-J., Heinkele, T., Jahnke, C. & Wolter, R. 2003. Characterisation of groundwater vulnerability. *Proceedings of 1st International Workshop on Aquifer Vulnerability and Risk, AVR'03*, Vol.1, Salamanca, Mexico: 189–201.

Zwahlen, F. (ed.) 2004. Vulnerability and risk mapping for the protection of carbonate (karst) aquifers. Final report (COST Action 620). European Commission, Directorate-General XII Science, Research and Development; Brussels, Luxemburg.

CHAPTER 3

Dense hydrogeological mapping as a basis for establishing groundwater vulnerability maps in Denmark

R. Thomsen & V. Søndergaard

Groundwater Department, Environmental Division, Aarhus County, Højbjerg, Denmark

ABSTRACT: The water supply in Denmark is based on high quality groundwater, thus obviating the need for complex and expensive purification. Contamination from urban development and agricultural sources increasingly threatens the groundwater resource, however. In 1995, the Government thus launched a 10-point plan to improve groundwater protection, and in 1998 further decided to instigate spatially dense hydrogeological mapping of the groundwater resources within the 37% of Denmark designated as particularly valuable water abstraction areas. The objective of the mapping, which is based on new geophysical surveys, survey drilling, water sampling, hydrological modelling, etc., is to prepare groundwater vulnerability maps. These maps will be used to for establish site-specific groundwater protection zones to prevent groundwater contamination from urban development and agricultural sources. This paper reviews the Danish strategy to protect the groundwater resource and introduces the new geophysical methods developed to supply inexpensive data needed for hydrogeological and groundwater vulnerability maps.

1   INTRODUCTION

In 1995, increasing problems with water quality in Denmark due to urban development and contamination from agricultural sources led the Minister for the Environment to approve a 10-point plan to improve groundwater protection. One of the major initiatives was that the Counties (the regional authorities) should draw up new water resource protection plans. By the end of 1997, the 14 County Councils had classified the country into three types of groundwater abstraction area: Particularly valuable, valuable and less valuable water abstraction areas (Figure 1).

In July 1998, the Danish Parliament adopted an ambitious plan to significantly intensify hydrogeological investigation to facilitate protection of the groundwater resource in order to meet future water supply challenges. Parliament decided that in addition to being responsible for water resource planning, the 14 County Councils should also be responsible for ensuring spatially dense mapping and hydrological modelling of the water resources as a basis for establishing site-specific groundwater protection zones.

The mapping and planning work is to be carried out over a 10-yr period and encompasses all parts of Denmark classified as particularly valuable water abstraction areas. Together, these cover almost 16,000 km^2 or 37% of the total area of the country. The total cost of this spatially dense mapping and planning work is estimated at around DKK 920 million, equivalent to 120 million euro (2002 prices) or 7,500 euro per km^2 for a programme encompassing geophysical

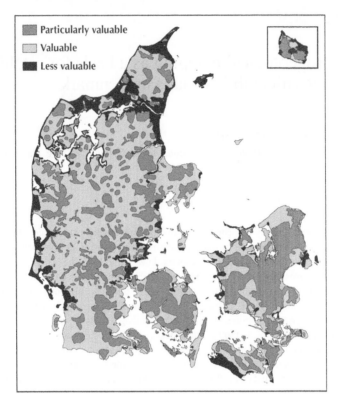

Figure 1.   Map of Aarhus County indicating particularly valuable, valuable and less valuable groundwater abstraction areas.

profiling every 250 m, survey drilling every 4 km², water sampling and hydrological modelling. During the 10-year mapping and planning period, Danish consumers have to pay the County Councils a 0.02 euro surcharge per m³ of water consumed, i.e. about 4 euro per family per year.

2   SITE-SPECIFIC GROUNDWATER PROTECTION STRATEGY AND ACTION PLAN

Groundwater protection in Denmark is based on the assumption that the physical environment provides some degree of protection against anthropogenic pressures, especially as regards contaminants entering the subsurface environment. The fundamental concept of site-specific groundwater protection zones is that some areas are more vulnerable to groundwater contamination than others. The goal is thus the subdivision of a given area according to the different potential of the various subareas as regards specific purposes and uses.

The Danish site-specific groundwater protection strategy is based on three steps (Thomsen et al. 2004):

1. Spatially dense hydrogeological mapping based on new geophysical surveys, survey drilling, water sampling, hydrological modelling, etc. aimed at facilitating the establishment

of site-specific protection zones. These zones are directed at both point sources and diffuse sources of contamination within the whole groundwater recharge area and shall supplement the traditional protection zones around the wells. The vulnerability is interpreted in relation to the local hydrological and chemical conditions.
2. Mapping and assessment of all past, present and possible, future sources of contamination – both point-source and diffuse.
3. Preparation and evaluation of an action plan stipulating politically determined regulations for future land use within the site-specific groundwater protection zones. The action plan has to be evaluated through a public planning process with a high degree of transparency and public participation. Moreover, it must include a timetable for implementation and a description of who is responsible for implementing the plan. The protection zones and guidelines will be used to prevent groundwater contamination from urban development and agricultural activities and for planning the remediation of contaminated sites.

The establishment of protection zones of this type imposes demanding requirements as to mapping of the water resources because the restrictions associated with the zones have to be set at property level. The Quaternary geology in Denmark is very complex, and the existing geological maps are largely based on geological information from boreholes. As a consequence, the maps are not sufficiently detailed and precise to enable delineation of the new protection zones.

3 THE DANISH HYDROLOGICAL SETTING

Denmark occupies a total area of some 43,000 km^2. The country consists of mainland Jutland (30,000 km^2), which is contiguous with Europe, and nearly 500 islands, of which more than 200 are inhabited. Denmark has been continuously populated and cultivated for over 3,500 years. Agriculture is one of the most important industries, and dominates the landscape. Most of the country consists of Quaternary deposits overlying Cretaceous chalk, limestone and Tertiary sand and clay. The topography is low-lying, reaching a maximum of 172 m above sea level. The combination of low topography and widespread consolidated and unconsolidated aquifers ensures a plentiful and easily accessible water resource. Groundwater recharge averages 100 mm per year, but can vary from 50 to 350 mm. Currently, approx. 800 million m^3 of water are abstracted annually. Household consumption by the 5.35 million inhabitants amounts to approx. 250 million m^3 per year. Of this, 99% derives from groundwater.

Groundwater quality in Denmark is generally good, thus obviating the need for complex and expensive water purification. The drinking water is not chlorinated and is of bottle water quality at the tap.

Spatially dense geophysical mapping in and around Aarhus County during the period 1994–97 has revealed that traditional mapping based solely on borehole data is too inaccurate for establishing site-specific groundwater protection zones and regulating land use within them. According to the Danish EPA guidelines, geophysical methods are expected to play an important role in hydrogeological mapping in Denmark. Experience with the application of geophysical methods to the mapping of large-scale geological structures around the city of Aarhus (see later) has been of great value. In recent years, new geophysical mapping methods have been developed through a collaborative effort by Aarhus County, Public Utilities of Aarhus and the University of Aarhus. The aim was to upgrade and rationalize fieldwork as well as to standardize the use of the mapping methods in a region. These new methods are very important tools for carrying out the spatially dense mapping needed to determine the extent and vulnerability and water quality of Danish aquifers as the basis for delineating protection zones. The geophysical mapping methods are described below.

4   MAPPING OF AQUIFER VULNERABILITY

Danish experience is that degradation of substances that can contaminate the groundwater is considerable in the plough layer, but that the thickness and the composition of the clay layers overlying the aquifers are the most important geological factors as regards protection of the groundwater against the types of contaminant that cause most problems in Denmark. It is therefore important to carry out spatially dense mapping of clay thickness and of the distribution and variation of sandy areas (windows) in the soil layers overlying the aquifers. The dense geophysical mapping is necessary because the complex geology makes it impossible to produce sufficiently detailed maps without geophysical data as there are usually only very few boreholes per km^2, and the majority of the boreholes are not particularly deep and not always well described.

As regards vulnerability mapping, the aim of the geophysical mapping is to determine the total clay thickness within the upper approx. 30 m b.g.s., to delimit any sandy windows present in the upper clay layers and to correlate the data with the information from the boreholes. In Denmark it has long been tradition to primarily base mapping of water resources on geological interpretation of boreholes, and geophysics were only normally used when it was in accordance with a previously established geological description.

Surface mapping with the new geophysical methods combined with better interpretation programmes has shown that it is time to do away with the old way of using geophysics. The quality with which water boreholes are described in Denmark varies considerably. Firstly, the choice of drilling method considerably influences the quality of the borehole samples that can be collected, and secondly, borehole drillers have been required to have an education encompassing at least a minimum of geological knowledge only since 2002. This means that a large number of borehole drillers are really good at drilling boreholes rapidly and cheaply, but unfortunately also that they are less good at collecting geological information of reasonable quality. The borehole driller now has to send borehole samples to the national archive of borehole data at the Geological Survey of Denmark and Greenland for sample description, but many samples were previously only described by the borehole driller himself. It is now accepted in Denmark, though, that borehole data has to be viewed critically. In the future, therefore, geological interpretation must be based to a greater extent on all the available information, with the weight accorded to it being to a greater extent determined by its quality.

By carrying out measurements with pulled array continuous electrical profiling it has been possible to map contiguous variations within the upper approx. 30 m b.g.s. that it would not be possible to map on the basis of borehole data alone. Based on combined interpretation of the geophysical measurements and the borehole information it is possible to produce accurate maps of the total clay thickness within the upper approx. 30 m b.g.s. Experience from investigations of water quality in particular has shown that if the total thickness of the clay layer within the upper 30 m b.g.s. comprises more than half of the total layer thickness at that depth interval, the groundwater is likely to be well protected against leaching of nitrogen and most of the other contaminants that typically pose problems in Denmark. When the total clay layer thickness exceeds 15 m, it obviates the problem associated with fissures in the clay layer as such fissures have not been observed to penetrate to depths greater than 7–8 m. The clay thickness map is compared with the groundwater quality and the results obtained with new monitoring boreholes.

The geophysical surface mapping was previously carried out by using resistivity data from electrical profiling measurements using the Wenner configuration for 10-, 20- and 30-m electrode separations with manual electrode placement. That method is rather expensive and time-consuming, and hence was only used for investigating smaller areas. In the early 1990s, Aarhus University developed the pulled array continuous electrical profiling method (PACEP) for obtaining electrical resistivity data in the Wenner configuration for 10-, 20- and 30-m electrode separations. The array pulled by a small caterpillar enables spatially dense measurements to be made along profiles (Figure 2). The largest contiguous area in Denmark in which apparent resistivity has been measured (Wenner, a = 30 m) is also shown in Figure 2. The majority of the pilot mapping studies carried out since the early 1990s are included in this figure.

Since the 1990s, the method has been improved to provide measurements at eight different electrode separations ranging from 2 to 30 m, and is now known as the PACES method (pulled array continuous electrical sounding) with Pol-Pol configuration 2, 3, 4, 5 and 15 m as well as Wenner configuration for 10-, 20- and 30-m electrode separations. Having more electrodes, the new method provides better possibilities for interpreting in multiple layers. The acquisition of eight rather than three measurements at each location enables full, three-layer inversion of the data.

The further interpretation and data analysis are carried out using the programme package GGGWorkbench. The three G's represent Geophysics, Geology and GIS. Geophysical processing and interpretation is carried out in a GIS environment integrating geophysical, geological and geographical data. The GGGWorkbench represents the newest generation of geophysical software from The HydroGeophysics Group. (See the website: http://www.hgg.au.dk) The GGGWorkbench operates on GERDA as its internal database for geophysical data. As previously mentioned, GERDA is the national database for geophysical measurements. (See the website: http://gerda.geus.dk). This new tool enables far more comprehensive geology-related interpretation of the large amounts of geophysical data to be performed than has previously been possible.

5   INDEPENDENT VULNERABILITY INDICATORS

When determining aquifer vulnerability it is appropriate to use several mutually independent indicators. In this connection, borehole observations and interpretation of geophysical measurements can usually serve as independent indicators.

The pulled array geoelectrical measurement method can be used to calculate the total thickness of the clay cover (also called the geophysical clay thickness) in the upper approx.

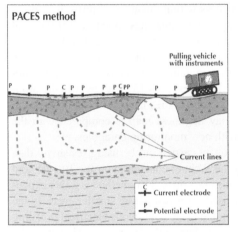

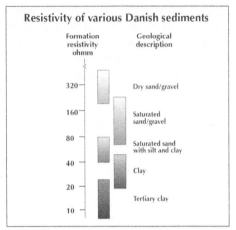

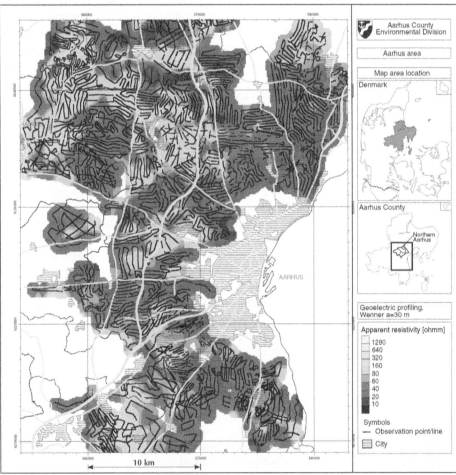

Figure 2.   Map of geoelectrical profile measurements (PACEP and PACES measurements) in an area around Aarhus. In areas with high electrical resistivity there is the risk that the groundwater is vulnerable.

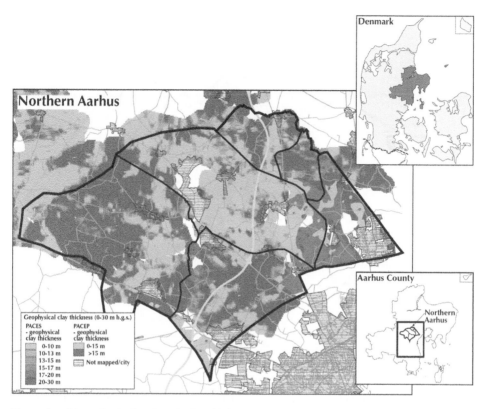

Figure 3. Map of calculated total clay thickness in the depth interval 0–30 m b.g.s. based on PACEP and PACES measurements (the geophysical clay thickness).

30 m b.g.s. (Figure 3). As mentioned earlier, an area is characterized as vulnerable if less than half of the upper 30 m b.g.s. consists of clay (layers with low resistivity, typically less than 50 ohm).

The geophysics-based calculations of clay thickness are thereafter compared with the observed clay thickness in boreholes. However, other borehole data exist that are just as interesting as the observed clay thickness with regard to determining vulnerability, i.e. the oxidation depth. This is determined from the colour description of the penetrated layers. The oxidation depth in a borehole is defined as the deepest depth to which reddish, red-brown, brownish, yellow-brown, yellowish and corresponding coloured soil layers are present, while the shades greyish, greenish and black indicate reduced conditions.

That geophysical measurements and borehole observations each contribute mutually independent information about the vulnerability of an area – especially nitrate vulnerability – is illustrated below using an example of vulnerability mapping in an area of approx. 160 km^2 located northeast of Aarhus.

A map of the observed oxidation depths in the boreholes in the area is shown in Figure 4. This is based on all 760 boreholes in the area, although information on the colour of the penetrated layers is lacking for about half of the boreholes, especially the oldest ones, which consequently do not contribute to the description of the oxidation depth. The oxidation

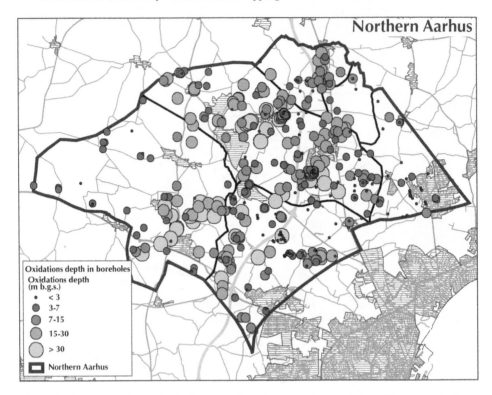

Figure 4.   Map of oxidation depths based on the colour description of the soil layers in the bore-holes in the area.

depth in the boreholes is determined without knowledge of the geophysical clay thickness in the area. As is apparent from Figure 4, it will be very difficult to describe the extent and boundaries of vulnerable areas with a high oxidation depth on the basis of borehole infor-mation alone.

A map comparing the information in Figures 3 and 4 is shown in Figure 5. The geo-physics-based calculations of clay thickness are shown here together with the observed oxidation depths in the boreholes. The map focuses on oxidation depths exceeding 7 m, as indicated by differently coloured circles. There is a very high degree of correlation between geophysical clay thickness of less than 15 m and high oxidation depths based on borehole information. This can be taken as an indication that the areas where the total clay thickness is less than 15 m within the upper 30 m b.g.s. are also the areas where the oxida-tion front lies deepest, and where substances such as nitrate will therefore be able to pene-trate deepest and thereby affect the water quality in the aquifers.

A map of the calculated total clay thickness at the depth interval 0–30 m b.g.s. based on PACES and PACEP measurements is shown in Figure 3. Areas where the calculated total clay thickness is less than 15 m, corresponding to the potentially vulnerable areas, are indi-cated in red. Note the irregular yet clear delimitation of the areas with a geophysical clay thickness of less than 15 m.

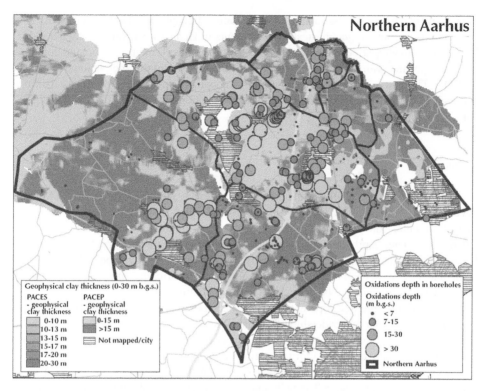

Figure 5.   Map comparing the geophysical clay thickness and information on the oxidation depths in boreholes focussing on oxidation depths exceeding 7 m. Note the correlation between small clay thickness and large oxidation depth, which indicates vulnerability in relation to nitrate, etc.

6   COMBINED INTERPRETATION OF DATA

The new mapping initiative includes setting up conceptual geological models and hydrological models for use in analysing the geological information and determining the water balance, the effect of climatic change, the extent of river basins, the capture zone of water supply wells and the contamination risk.

During the past four years, all the major water utilities in Denmark have gained experience in all the aspects of spatially dense hydrogeological mapping. The cooperation between the HydroGeophysics Group at Aarhus University and the Counties is very important for ensuring high quality in the collection and interpretation of geophysical data. The HydroGeophysics Group at Aarhus University works intensively to continually improve the measurement methods and data interpretation. The extensive measurement campaign has shown that geophysics is presently a cheap and effective method to fill in the geological picture between boreholes. The practical work of undertaking the geophysical measurements and interpreting the results has been carried out by private consultancy firms. The quality assurance procedures for the geophysical measurements and the programming of interpretation programmes are paid for by the Counties and the University. The work is carried out by the HydroGeophysics Group and made available free of charge to the consultants.

Measurement data have to be submitted to the national geophysics database GERDA. The cost of carrying out the geophysical measurements in Denmark has now reached a stable level. The following price examples encompass measurement, interpretation and comprehensive reporting with various theme maps, including reporting to the National database GERDA. The costs are based on mapping tasks encompassing an area of at least $10 \, km^2$, and include 25% Danish VAT. Providing these assumptions are met, PACES will cost 550 euro, with 4 lines needed per km^2.

7   SKYTEM

In 2003, a new transient electromagnetic measurement method termed SkyTEM was developed in Denmark. The SkyTEM method was developed specially for mapping geological structures down to depths that under Danish conditions correspond to around 150 m b.g.s. This depth interval is particularly interesting in large parts of Denmark because it is at these depths that one usually finds the Prequaternary buried valleys that contain a large proportion of the aquifers upon which the drinking water supply is based, and which therefore need to be protected against contamination.

The new, pioneering development has been made possible by new measurement equipment developed under the leadership of geophysicist Kurt Sørensen, University of Aarhus. The development of the apparatus is based on work over the past 10 years with a Canadian instrument originally developed for measuring ore deposits.

The helicopter used for the measurement campaigns flies at a height of approx. 60 m at a speed of 20–30 km per hour. The helicopter-based measurement method means that it is possible to perform measurements during a larger part of the year, while at the same time ensuring a higher mapping rate and hence improved effectiveness.

Based on experience to date with use of the method to map a total area of $200 \, km^2$ the results seem very promising with respect to assessment of that part of aquifer vulnerability associated with the thickness of the clay layers overlying the aquifers, even though the main objective when developing the new SkyTEM method was mapping of the location, extent and depth of the buried valleys and aquifers. Thus it seems that it may be possible to use one and the same method to collect information both about aquifer location and extent and about the vulnerability of the groundwater. In future it should thus be possible to intensify mapping of aquifers and their vulnerability, while concomitantly ensuring high mapping quality.

8   EXPERIENCE WITH GROUNDWATER PROTECTION IN AARHUS COUNTY

In Aarhus municipality, the total thickness of clay layers above the aquifers has been determined by combined interpretation of geophysical and geological data. Groundwater protection zones are now being established in the particularly valuable water abstraction areas on the basis of the spatially dense geophysical mapping and hydrological modelling. After negotiation with Aarhus County, Aarhus Municipality has accepted the importance of active protection of the groundwater resources as a basis for ensuring a future drinking water supply of good quality. All new urban development around Aarhus city must take groundwater interests into consideration, and urban development will not usually be permitted in areas where natural protection of the groundwater is poor. Moreover, plans to develop 1,000 ha of land designated for urban development have been abandoned in 2001 as a result of the new information. In 1997, Aarhus County Council decided that no new areas could be designated for urban development before spatially dense hydrogeological mapping of the areas in question had been carried out. It is estimated that the groundwater is protected by clay layers in 50% of Denmark.

9   CONCLUSION

It is expected that the site-specific groundwater protection zones currently being established to ensure the future water supply in Denmark will substantially influence future urban development and land use. It is thus important that the protection zones are based on spatially dense hydrogeological mapping encompassing geophysical and geochemical data. The geophysical mapping will be performed using PACES and SkyTEM. The site-specific protection maps designating the protection zones and associated regulation of land use will be used to prevent groundwater contamination from urban development and agricultural activities and for planning remediation of contaminated sites. Mapping and establishment of the groundwater protection zones will take place over a 10-year period at a total cost of around 120 million euros. During the 10-year period consumers will pay the County Councils a 0.02 euro surcharge per m^3 of water consumed, i.e. less than 4 euro per family per year. This ongoing Danish initiative to draw up spatially dense hydrogeological maps of the 37% of Denmark designated as particularly valuable water abstraction areas will have to be adjusted to the new EU Water Framework Directive as the latter encompasses all water bodies.

REFERENCE

Thomsen, R., Søndergaard, V. H. & Sørensen, K. I. 2004. Hydrogeological mapping as a basis for establishing site-specific groundwater protection zones in Denmark. *Hydrogeology Journal*, vol. 12: 550–562.

CHAPTER 4

The Polish concept of groundwater vulnerability mapping

S. Witczak, R. Duda & A. Zurek

Department of Hydrogeology and Water Protection, AGH – University of Science and Technology, Krakow, Poland

ABSTRACT: The paper presents the concept behind and the methodology used in the production of the 1:500,000 groundwater vulnerability map of Poland. The approach assumes that vertical transport time through the vadose zone is the most important factor of the risk. The map is composed of three sheets referring to different elements of hydrogeological systems. Sheet 1 – intrinsic vulnerability of shallow groundwater – includes classes of aquifer vulnerability evaluated on the basis of the vertical transport time of conservative contaminants, as well as the velocities and directions of their horizontal transport. The classification of shallow aquifer pollution vulnerability is based on the approach given by Foster et al. (2002). Sheet 2 shows the specific vulnerability of the shallow groundwater to nitrate leaching. Sheet 3 represents vulnerability of Major Groundwater Basins evaluated on the basis of the transport time of conservative contaminants from the recharge area on the land surface to the basin border.

1 INTRODUCTION

The concept of groundwater pollution vulnerability mapped at a scale of 1:500,000 is complex. The approach assumes a multilayer system with superposition of aquifers and simultaneously follows the requirements of current European (EWFD 2000, DGWP unpubl.) and Polish legislation. Preparation of the groundwater vulnerability map includes:

- assessment of the vulnerability of groundwater interacting with surface waters and those terrestrial ecosystems which status closely depends on the quantity and quality of groundwater i.e. wetlands, peat-bogs, parts of forest ecosystems,
- assessment of the vulnerability of aquifers essential for the provision of drinking water, i.e. the Major Groundwater Basins (MGWB). The following basic criteria are used to classify MGWBs: the presence of at least one well having a yield greater than $70\,m^3/hour$, total groundwater abstraction greater than $10,000\,m^3/day$, transmissivity greater than $240\,m^2/day$, and water quality of class I, i.e. very high quality (Kleczkowski ed. 1990). These criteria have been lowered in areas with a deficit of water, particularly in the southern Poland.

The final map will be only of a general and strategic importance. Any further protection planning with respect to watersheds requires detailed maps, the best being at the scale of 1:50,000 or larger.

2   DEFINITION OF GROUNDWATER VULNERABILITY

Groundwater vulnerability is a complicated issue and this is reflected in the various definitions and methodologies for its assessment that have been published. Methodological problems pertaining to vulnerability assessment depend on the complexity and variability of recharge and groundwater flow conditions in various hydrogeological media: porous, fissured-porous and fissured-karst. Evaluation of groundwater vulnerability can be carried out using a range of different methods (Aller et al. 1987, Foster 1987, Robins et al. 1994, Vrba & Zaporozec 1994, Holting et al. 1995, Doerfliger et al. 1999, Hannapel & Voight 1999, Gogu & Dassargues 2000, Daly et al. 2002, Foster et al. 2002, Zurek et al. 2002).

Groundwater pollution vulnerability is a natural property of a water-bearing system defining a risk of migration of harmful substances from the surface to the aquifer. Intrinsic vulnerability (also called natural vulnerability) is controlled exclusively by geological structure and hydrogeological conditions, while specific vulnerability includes, besides the former parameters, consideration of the type of a contaminant and the character of a contamination source (Vrba & Zaporozec 1994).

The methodology described below refers to the determination of the intrinsic (natural) vulnerability, referred to hereafter as "vulnerability". Depending on the assumed effects of selected contaminants or of land-use planning, it is possible to predict a specific vulnerability or prepare risk scenarios. To start with, the authors propose the delineation of the areas vulnerable to contamination with agriculture-related nitrate.

3   METHODOLOGICAL ASSUMPTIONS

No unified methodology of vulnerability assessment has been accepted in Poland, although the problem has been presented in several papers (e.g. Kleczkowski ed. 1990, Witkowski et al. 2002). For this reason, when developing the current concept for the groundwater vulnerability map of Poland, the authors based it not only on their own experience (e.g. Witczak & Zurek 2000, Zurek et al. 2002) but also made use of concepts published abroad.

In selecting the methodology for the vulnerability assessment and map preparation, particular attention was focussed on elements of mutual interaction of quality and quantity between groundwater and surface waters within the watershed. The following three essential assumptions have been made:

- The flow of groundwater is 3-D (Winter et al. 1998, Alley et al. 1999). Therefore, it has been necessary to consider multilayer systems and superposition of aquifers, and to propose the construction of the map in the form of separate sheets.
- The quality of surface waters depends on the quality of groundwater. For the majority of the year groundwater baseflow forms 80–90% of the surface flow (Duda et al. 1996). As a consequence, groundwater quality controls the quality of surface waters and the degree of dilution of effluents and wastewaters.

Surface waters are recharged not only by the groundwater coming from the useful groundwater horizons or highly water-bearing horizons, but also by groundwater flowing from shallow systems not belonging to the category of the useful groundwater horizons. As a consequence, these shallow aquifers of lesser capacity are usually omitted from the groundwater mapping, although the real outflow from such shallow systems may be up to

30% of the total stream-flow (Prazak et al. 2001, Witczak et al. 2003). European Community policy has enforced a new approach to areas with low groundwater capacity (Dillon & Simmers 1998, Identification... 2003) because vulnerability maps provide one of the fundamental tools for land-use planning and construction of programs for the prevention of contamination of surface waters, including their eutrophication. Thus, the first sheet of the vulnerability map concerns the shallow groundwater interacting with surface waters and terrestrial ecosystems.

The travel time of groundwater is long-term, several tens of years on the average. Long-term transport of contaminants in groundwater results, in turn, in a retarded reaction of surface waters to quality changes in groundwater baseflow. As a result, several years or even several tens of years following cessation of discharge, the concentration of the contaminants carried in the groundwater baseflow into rivers may only be halved (Duda et al. 1996). It should be noted that this retardation refers only to conservative contaminants, while absorbed contaminants will migrate for much longer periods. Therefore, the transport time of conservative contaminants is one of the essential elements of the vulnerability assessment and mapping.

In the assessment of groundwater vulnerability to contamination from agriculture-related nitrate the fact that the current nitrogen load carried by groundwater baseflow is a result of intensive fertilization in the past should be taken into account. In this situation, a proper approach to delineate zones sensitive to nitrate contamination will have to include a correlation between retardation associated with a travel time, i.e. with the age of groundwater, and concentrations of nitrate in groundwater observed during current monitoring (Stockmarr 2001).

4 VULNERABILITY ASSESSMENT AND PREPARATION OF THE MAP

The proposed methodology of vulnerability assessment and mapping is presented as the description of individual information layers with their cartographic visualization on the map. The map is composed of three sheets of equal importance, representing the vulnerability assessment:

- groundwater interacting with surface water and terrestrial ecosystems dependent on the shallow groundwater;
- groundwater that can be contaminated with nitrates from agricultural sources;
- Major Groundwater Basins (MGWBs).

These information layers have been selected through a consideration of the possibilities of their being processed using GIS to obtain synthetic information on vulnerability. The map proposed is not only a graphic visualization in the form of four sheets but primarily represents a database with all the information layers selected. The database can easily be extended in the future to include any additional data required for specific scenarios/situations.

4.1 *Intrinsic vulnerability of groundwater interacting with surface waters and terrestrial ecosystems dependent on the shallow groundwater – sheet 1*

This is the main information layer. The classes of aquifer pollution vulnerability with their general definitions, based on Foster et al. (2002), and the vertical travel times of conservative

Table 1.   Classes of aquifer contamination vulnerability [after Foster et al. (2002), modified by the authors] and vertical travel times of conservative contaminants to aquifer.

Vulnerability class	Definition	Vertical travel time to aquifer [years]	Colour on the map
Very high	Aquifer vulnerable to most water pollutants with rapid impact in many pollution scenarios	<5	Red-orange
High	Aquifer vulnerable to many pollutants, except those strongly absorbed or readily transformed, in many pollution scenarios	5–25	Pink
Moderate	Aquifer vulnerable to some pollutants, but only when continuously discharged or leached	25–50	Yellow
Low	Aquifer only vulnerable to conservative pollutants in the long term when continuously and widely discharged or leached	50–100	Light olive-green
Very low	Aquifer confining beds present with no significant vertical groundwater leakage	>100	Olive-green

contaminants (Table 1) are presented in colour following the IAH scale (Vrba & Zaporozec 1994).

In any assessment of intrinsic vulnerability to infiltrating contaminants, their vertical transport time through the vadose zone is the most important factor. It depends mainly on the thickness and hydrogeological properties of the strata of the vadose zone. For shallow, and therefore the most vulnerable groundwater, the thickness of the vadose zone is measured as the depth to the water table. The mass transport time (t_a) may be described by the following Equation 1, assuming the piston flow model:

$$t_a = \frac{m_a \cdot w_o}{P \cdot \varpi_i} \tag{1}$$

where m_a = thickness of the vadose zone [L]; w_o = average volumetric water content of the strata in the vadose zone [−]; P = mean annual precipitation [L/T] and ω_i = effective infiltration coefficient [−].

For typical grain-size distributions in Polish soils, values of the volumetric water contents and the effective infiltration coefficients for near-surface strata are given in Table 2. For areas with significant surface gradient, the assumed effective infiltration coefficients should be corrected due to the increased surface runoff.

In considering the groundwater vulnerability to contamination, due attention must be paid to aquifers occurring in fissured-karst regions (Doerfliger et al. 1999, Daly et al. 2002, Witkowski et al. 2002). In areas where such aquifers out crop recharge is very intensive, reaching even up to 50% of precipitation, and is accompanied by significant groundwater flow velocities. For this reason, most outcrops of fissured-karst aquifers should be considered as very highly or highly vulnerable to contamination.

Table 2. Soil protective capacity and the vertical transport times of conservative contaminants through 1 m of the soil profile

Soil protective capacity	Grain-size distribution groups of soils	Effective infiltration coefficient [−]	Volumetric water content [−]	Vertical travel time through 1 m of the soil profile* [years]
Very weak	Sand: loose, loose silty, weakly loamy weakly silty	0.17–0.25	0.11–0.14	0.7–1.4
Weak	Sand: light loamy light silty; sandy silt	0.13–0.17	0.14–0.21	1.4–2.7
Moderate	Loam: light and silty, medium and silty; loamy silt; loess	0.08–0.13	0.21–0.27	2.7–5.6
Good	Loam: heavy and silty; clayey silt and silty clay	0.05–0.08	0.27–0.46	5.6–15.3

* Calculated for the mean annual precipitation equal to 600 mm/year, assuming the piston flow model.

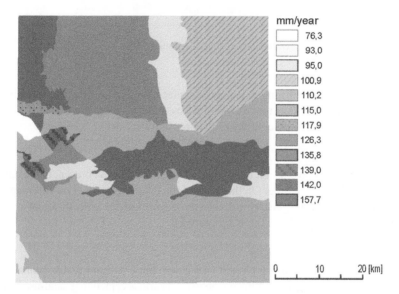

Figure 1. Recharge spatial distribution.

4.1.1 *Intensity of groundwater recharging by infiltration*

This information layer is not plotted on the final map but is present only in the database. It is used in GIS data processing. The recharge can be evaluated on the basis of renewable water resources per unit area. For instance, a recharge of 100 mm/year corresponds to 3.17 l/s/km². The recharge can be interpolated on the basis of characteristics of the renewable water resources from the Hydrological Map of Poland (HMP) 1:50,000 (Paczynski et al. 1999), but only for those hydrogeological units which are included the first aquifer below the land surface, the so called mainu usable aquifer (Figure 1). In other areas and also in the areas declared as "non-water-bearing" or "outside the main useful horizon", the renewable resources are interpolated on the basis of additional hydrological data.

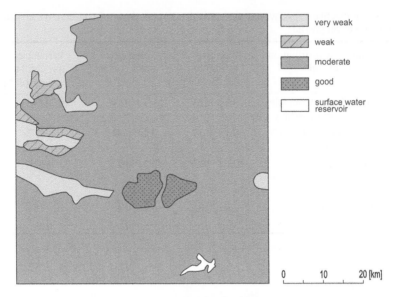

Figure 2.   Soil profile protective capacity spatial distribution.

Such a rather complicated approach, may be replaced in the future by the diversification of effective infiltration within the area studied by superposition of current information pertaining to soil character, topography of the area and land-use.

4.1.2   *Characteristics of soil protective properties*

Again, this information layer is not being plotted on the map but is present in the database. It is used in GIS data processing. On the basis of the details contained on soil maps available at various scales, four generalized classes of soil profiles with different protective properties can be distinguished (Table 2, Figure 2).

4.1.3   *Depth to the shallow aquifer*

This information is also not being plotted on the map but is present in the database and is used in GIS data processing, depending on the risk scenario accepted (e.g. land-use character, the presence of a specific contaminant). It represents the depth from the terrain surface to the first groundwater aquifer, the depth being expressed on the contour map in five ranges: <2 m, 2–5 m, 5–10 m, 10–20 m and >20 m (Figure 3) Such aquifers are important for terrestrial ecosystems dependent on groundwater (wetlands, sensitive forest habitats, etc.) and their vulnerability assessment is required by the EWFD (2000).

4.1.4   *Times and directions of the groundwater flow within the first aquifer from the surface*

This information layer will be marked on the map in the form of arrows. The length and description of the arrows express the horizontal transport time of conservative contaminants in which they cover a marked distance (Table 3, Figure 4). The proposed arrow length of 6 mm on the map at 1:500,000 represents a transport distance of 3 km. The value describing the arrow indicates the duration of transport time on the arrow distance (in years).

Table 3. Transport times and velocities of conservative contaminants within shallow aquifers.

Symbol on the map	Transport time of conservative substances [years]	Real flow velocity [m/year]	Transport velocity of conservative substances
$\underset{\rightarrow}{10}$	10	>300	Very fast
$\underset{\rightarrow}{20}$	20	100–300	Fast
$\underset{\rightarrow}{50}$	50	30–100	Medium fast
$\underset{\rightarrow}{100}$	100	<30	Slow and very slow

This symbol also helps assess the retardation of the surface water reaction to the change of the contamination load introduced into groundwater. The values of the arrows are calculated for characteristic points of the watershed area with representative hydrogeological parameters (hydraulic conductivity, active porosity) using the following sources: groundwater contour map and/or the topographic map, hydrogeological database, and regional reports.

4.1.5 *Groundwater quality of the shallow aquifer*
The general groundwater quality class is point-visualized in the form of graphic symbol, based on data from the state and regional groundwater quality monitoring networks and the results of measurements carried out during the preparation of the HMP 1:50,000. Five quality classes of groundwater are distinguished (Kazimierski & Macioszczyk 2003). Classes I, II and III appoint the good chemical status, classes IV and V appoint the poor chemical status of groundwater (Figure 4):

- class I = waters of very good quality; natural chemical composition of waters; none of quality determinants exceeds maximum permissible levels for drinking water;
- class II = waters of good quality; natural chemical composition of waters; only Fe and Mn exceed maximum permissible levels for drinking waters;
- class III = waters of sufficient quality; natural chemical composition of waters; Fe, Mn and five other non-toxic determinants exceed maximum permissible levels for drinking waters; simple water treatment required;
- class IV = waters of inferior quality; up to ten non-toxic and toxic quality determinants exceed maximum permissible levels for drinking waters; advanced water treatment required;
- class V = waters of bad quality; anthropogenic and/or geogenic contamination in the degree excluding utilization.

4.1.6 *Terrestrial ecosystems whose status depends on shallow groundwater*
The two following terrestrial ecosystems have been distinguished and marked with hatching on the basis of the CORINE land cover and land-use map (CLC 1995):

- forests situated in the zones, where the first water table is situated not deeper than 2 m from the land surface; the soils and habitats of such areas are strongly affected by groundwater;
- wetlands and peat-bogs (after Nawrocki & Madgwick 1999).

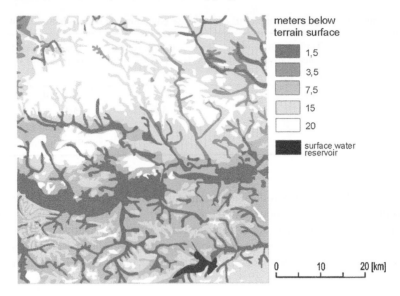

Figure 3.   Average depth to the shallow aquifer spatial distribution.

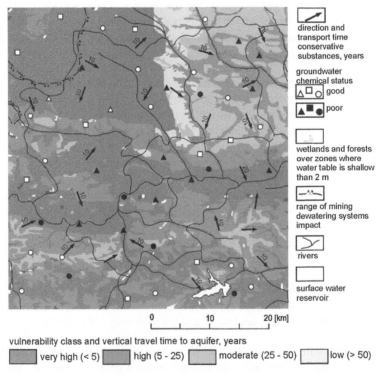

Figure 4.   Groundwater vulnerability map. Sheet 1 – intrinsic vulnerability of shallow groundwater interacting with terrestrial ecosystems.

4.1.7 *The risk of qualitative and quantitative changes of groundwater and surface waters mutually affecting each other*

This information layer is presented in linear form on the map and includes the following:

- river sections with infiltrating character, i.e. those in the natural situation, in the areas of infiltration intakes, within depression cones of groundwater intakes, within depression cones of mining dewatering systems;
- the range of impact of mining dewatering systems.

4.2 Specific vulnerability of the shallow groundwater to nitrates leaching – sheet 2

In the case of a nitrate-caused risk, protection against eutrophication of surface waters is a main objective. Surface waters are vulnerable to an input of nitrate from groundwater baseflow and from surface runoff. Nitrate originates from natural mineralization of organic matter in soil and, directly, from fertilization with mineral and organic compounds.

Surface runoff carries the nitrogen compounds contained in washed out fertilizers and the organic substance removed from soil during weathering into rivers. The determination of nitrogen loads attributable to surface runoff is methodologically difficult. Often the surface runoff and the groundwater baseflow are treated jointly as the net precipitation (i.e. the difference between the values of precipitation and evapotranspiration) and, in general terms, expressed as the value of the total runoff to rivers. However, the nitrate contained in surface runoff represent the load that was washed out from the arable land several days earlier, while the nitrate reaching surface waters from groundwater baseflow are an effect of fertilization from a distant past, several years or even several tens of years before present. As the ages of the waters from these two forms of outflow are so different, the load of nitrate reaching surface waters as a result of surface runoff should be calculated separately. The value of the surface runoff is evaluated as the difference between the value of the net precipitation and the groundwater baseflow. The net precipitation values, or the total runoff, are available from hydrological reports. There are also empirical equations for the estimation of the total runoff, depending mainly on topography of the watershed. The measured values of nitrogen load due to surface runoff are also available (Knisel & Davis 1999).

4.2.1 *Nitrates-vulnerable zones (NVZ) of groundwater*

This is the main information layer that will appear on sheet 2 and it will be superficially marked with colours.

Infiltrating waters, which seep through the soil profile, contaminate groundwater and leach nitrogen compounds, mainly nitrates. In accordance with the current priorities of the EC, sheet 2 will indicate the zones vulnerable to these quality changes of groundwater that are attributable to agriculture-related nitrate.

In the identification of NVZs, the authors have assumed, first of all, the vertical transport time through the vadose zone to be the most important factor of the risk. Applying this approach, the NVZs occur only in the areas of very high, high or moderate intrinsic vulnerability (Table 1). The final aquifer vulnerability to agriculture-related nitrate is a compilation of very high or high or moderate intrinsic vulnerability resulting from the vertical transport time through the vadose zone and a calculated concentration of nitrate in groundwater that is evaluated in geochemical models.

The vertical transport time depends mainly on the thickness and hydrogeological properties of the strata of the vadose zone. For the most shallow and most vulnerable groundwater, the thickness of the vadose zone is measured as the depth to the water table. The vertical transport time (t_a) is evaluated by the Equation 1 described earlier.

The concentration of nitrates in groundwater is evaluated on the basis of the model load of nitrogen carried into groundwater due to the fertilizer used and a mean annual net recharge of groundwater. In this approach, a concentration of nitrate in infiltrating water is calculated according to the Equation 2:

$$C_{NO_3} = \frac{L_N \cdot k_N}{P \cdot \varpi_i} \cdot 443 \qquad (2)$$

where C_{NO_3} = concentration of nitrates in infiltrating water [mg/dm³]; L_N = load of nitrogen in the fertilizer used [kg/hectare/year]; k_N = fraction of nitrogen leached to groundwater [−]; P = mean annual precipitation [mm/year]; ϖ_i = effective infiltration coefficient [−] and 443 = a constant resulting from recalculation of the nitrogen load leached to groundwater into the concentration of nitrate in infiltrating water.

Quantitative modeling of the load of the nitrates leached to groundwater (L_N*k_N) due to fertilization can be done utilizing numerical models, for instance GLEAMS (Knisel & Davis 1999), DAISY (Refsgaard et al. 1999) and STONE (www.riza.nl/projecten_nl).

Application of liquid fertilizers (mainly liquid manure) and sprinkling of plants only slightly increase infiltration and accelerate the process of nitrate leaching. Fertilization with the load of liquid manure at the advised level 45 m³/hectare/year will increase average infiltration, which is approximately 100 mm/year (1,000 m³/hectare/year), at around 4–5% only.

It should be noticed that because of the general map scale (1:500,000), the approach to evaluation of the nitrate load leached to groundwater is very simplified. More sophisticated scenarios should be used for a vulnerability map at a scale of 1:50,000.

Nitrate from groundwater reaches surface waters through baseflow. A long groundwater travel time results in a retarded surface water response to changes in contamination recharged to groundwater. The present-day concentration of nitrate in groundwater is a response to the intensity of past application of fertilisers at the surface – possibly in the late 1970s to 80s. Therefore, delineation of areas vulnerable to nitrate contamination based on statistical analysis of current nitrogen loads at the land surface compared with the current observed quality of shallow groundwater is not correct.

A hydrogeological approach should introduce a correlation that takes into account retardation resulting from the age of groundwater. Stockmarr (2001) presented such a correlation between historic periods of nitrogen fertilization in Denmark and concentrations of nitrates in the groundwater observed during a 50-year monitoring.

The evaluation of the amount of nitrates leached from the soil does not take into account the hazard represented by non-fertilized areas left fallow, nor that by the areas where a drastic change of planting took place, i.e. from permanent crops (e.g. green crops) to arable land. In such areas, because of mobilization of an enormous load of soil nitrogen, a dramatic increase of the amount of nitrates leached into groundwater is observed (Howard 1985, Juergens-Gschwind 1989, Worall & Burt 1999, Zurek 2002).

Nitrates are the most mobile and conservative of nitrogen compounds in the soil-water environment. Denitrification is the only process that can considerably lower the groundwater content of nitrate. This process can occur in the soil profile, vadose zone or the aquifer itself in the presence of organic substances and sulphides or oxides of bivalent iron. For instance, a Belgian concept (Eppinger & Walraevens 2002) of determination of the zones vulnerable to leaching of nitrates from agricultural sources links the borders of such zones with the so-called rock reductive storage. This coefficient is expressed as the content of organic substance typical of Quaternary and Tertiary deposits (originated from terrestrial sedimentation) and pyrite (mainly in deposits of marine origin) in the soil profile. However, at the inflow of a significant nitrate load, an effect of their denitrification that is positive for groundwater quality may also create a risk caused by considerable amounts of sulphate formed during oxidation of sulphides (Broers 2004).

4.2.2 *Times and directions of the groundwater flow within the shallow aquifer*

The directions and the times of groundwater flow between the recharge areas where the contaminant load is introduced and the drainage areas that are vulnerable to eutrophization by surface waters are well visualized by a system of arrows (Table 3). The length and description (years) of the arrows characterize the time in which a conservative contaminant migrates over a marked distance, similarly to described in part 4.1.4.

4.2.3 *Groundwater quality*

The concentration of nitrates as a current groundwater quality indicator will be point-visualized in the form of graphic symbols based on data from the groundwater monitoring system and the results of measurements carried out during the preparation of the HMP 1:50,000. The groundwater quality classes (Kazimierski & Macioszczyk 2003) are as follows: <10 mg NO_3/dm^3 = I class; 10–25 mg NO_3/dm^3 = II class; 25–50 mg NO_3/dm^3 = III class; 50–100 mg NO_3/dm^3 = IV class; >100 mg NO_3/dm^3 = V class. Into the classes I, II and III have good chemical status, and classes IV and V are of inferior status.

4.3 *Vulnerability of the Major Groundwater Basins – sheet 3*

This sheet will be marked with colours on the basis of the intrinsic vulnerability of the basin to contamination from the surface. The level of the intrinsic vulnerability of the MGWB is expressed as the total of the vertical travel time of conservative contaminants from the surface to the basin and the horizontal transport time of these contaminants to the basin borders within the limits of its watershed area. Accepting the piston flow model of migration, conservative contaminants migrate in groundwater at the rate of the intrinsic flow velocity of groundwater.

4.3.1 *The vulnerability of the MGWBs and their recharge areas*

The vulnerability of the MGWBs and their recharge areas can be classified as follows:

- extremely vulnerable and highly vulnerable = travel time shorter than 5 years; the basin and its recharge area are very highly vulnerable ($t < 2$ years) or highly vulnerable ($2 < t < 5$ years) and require extreme protection, the so called Maximum Protection Area (MPA);
- moderately vulnerable = travel time 5–25 years; the basin and its recharge area are moderately vulnerable and require high protection, the so called High Protection Area (HPA);
- low and very low vulnerable = travel time longer than 25 years; the basin is of low vulnerability ($25 < t < 100$ years) or very low vulnerability ($t > 100$ years).

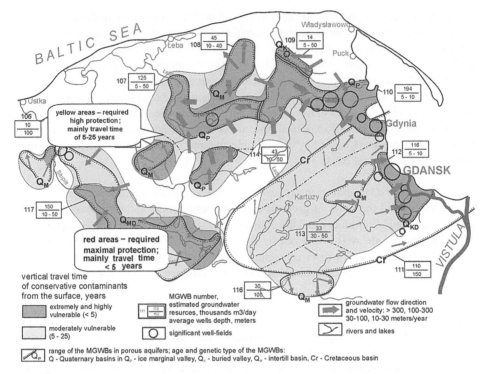

Figure 5. Groundwater vulnerability map. Sheet 3 – intrinsic vulnerability of the Major Groundwater Basins (MGWB) and their recharge areas (after Kleczkowski ed. 1990).

The borders of the protection areas and the areas of the MGWBs for which detailed reports at 1:50,000 have not yet been prepared are taken from the MGWBs map (Kleczkowski ed. 1990). For the basins for which detailed reports are available, the borders of both the basin themselves and their protection areas are taken from these reports. The MPA (the total area having travel times shorter than 5 years) and the HPA (the total area having travel times shorter than 25 years) are equivalents of highly vulnerable and moderately vulnerable MGWB areas (Figure 5). In some cases, the borders of these vulnerability classes have been extended beyond the hydrogeological borders of the MGWB, into their recharge areas, to obtain the total transport times of conservative contaminant from the surface to the borders of the MGWB equal to 25 years.

This sheet, being a part of the vulnerability map, is to represent one of the essential factors in updating land-use plans. The location of potential, industrial and agricultural contamination sources within the protection areas (MPA and HPA) from which contaminants may migrate from the surface to the basin borders in less than 25 years should be prohibited.

4.3.2   *The qualitative status of groundwater*
The qualitative status of groundwater in the MGWBs will be point-visualized with graphic symbols on the basis of the quality classification of groundwater, similarly to described in part 4.1.5.

5 SUMMARY

Maps of both intrinsic (natural) and specific vulnerabilities of aquifers are required for implementation of the Framework Water Directive. The approach of Foster et al. (2002) to the vulnerability assessment, modified by the introduction of the total migration time of contaminants into groundwater proposed by the authors, may become a basis in preparation of such maps. A vulnerability map fulfilling the requirements of the Directive should be composed of several sheets. The data acquired during preparation of the map in the version proposed in this paper should form a nucleus of a wider hydrogeological and environmental database that can be enlarged in the future during preparation of maps with different risk scenarios or new maps of groundwater vulnerability.

ACKNOWLEDGEMENTS

The investigations presented are part of statutory research of the Dept. of Hydrogeology and Water Protection, AGH-University of Science and Technology, Krakow, Poland, financially supported by KBN grant no 11.11.140.139.

The authors thank Dr. Joanna Karlikowska for her help in the GIS and preparation of figures.

REFERENCES

Aller, L., Bennett, T., Lehr, J.H., Petty, R.J. & Hackett, G. 1987. DRASTIC: a standardized system for evaluating ground water pollution potential using hydrogeological settings. EPA Reports, 600/2-87-035, Washington DC, USA.

Alley, W.M., Reilly, T.E. & Franke, O.L. 1999. Sustainability of ground-water resources. *US Geol. Survey Circular*, 1186. Denver, Colorado: p. 79.

Broers, H.P. 2004. Nitrate reduction and pyrite oxidation in the Netherlands. In: *Nitrates in Groundwater; IAH Hydrogeology Selected Papers,* 5, (Razowska-Jaworek & Sadurski eds.), Rotterdam: Balkema.

CLC. 1995. CORINE Land Cover. EEA Reports about Europe's environment: Commission of the European Communities, Copenhagen: p. 163.

Daly, D., Dassargues, A., Drew, D., Dunne, S., Goldscheider, N., Neale, S., Popescu, I.C. & Zwahlen, F. 2002. Main concepts of the "European approach" to karst-groundwater-vulnerability assessment and mapping. *Hydrogeology Journal* 10: 340–345.

DGWP. unpubl. *Proposal of Directive of the European Parliament and of the Council on the protection of groundwater against pollution.* European Commission, COM(2003) 550 Final draft. Brussels.

Dillon, P. & Simmers, I. 1998. Shallow groundwater systems. *IAH International Contributions to Hydrogeology* 18, Rotterdam: Balkema.

Doerfliger, N., Jeannin, P.Y. & Zwahlen, F. 1999. Water vulnerability assessment in karst environments: A new method of defining protection areas using a multi-attribute approach and GIS tools (EPIK method). *Environmental Geology* 39(2): 165–176.

Duda, R., Witczak, S. & Bednarczyk, S. 1996. Regional groundwater quality monitoring as a tool for the base flow quality modeling of the Upper Vistula River Basin (SE Poland). In: *Vol. of Poster Papers; IAHS Inter. Conf. on Application of Geographic Information Systems in hydrology and water resources management* (H. Holzmann & H.P. Nachtnebel eds.), Vienna: 91–97.

Duda, R., Witczak, S. & Zurek, A. 2003. The concept of groundwater vulnerability map. In: *Wspolczesne Problemy Hydrogeologii*; v. XI, (Piekarek-Jankowska & Jaworska-Szulc eds.), Gdansk Univ. of Technology, Gdansk (in Polish): 269–278.

Eppinger, R. & Walraevens, K. 2002. Spatial distribution of nitrate in aquifers controlled by a variable reactivity system. In: *Nitrates in Groundwater; IAH Hydrogeology Selected Papers*, 5 (Razowska-Jaworek & Sadurski eds.), Rotterdam: Balkema.

EWFD. 2000. Directive 2000/60/EC of the European Parliament and of the Council of 23 October 2000 establishing a framework for Community action in filed of water policy. *Official Jour. of the European Com.*, L327. Brussels.

Foster, S. 1987. Fundamental concepts in aquifer vulnerability, pollution risk and protection strategy. In: *Vulnerability of soil and groundwater to pollutants*; 38 (van Duijvenboden & H.G. van Waegeningh eds.), TNO Commission on Hydro Res, Proc. and Inform., Hague: 69–86.

Foster, S., Hirata, R., Gomes, D., D'Elia, M. & Paris, M. 2002. Groundwater quality protection. A guide for water utilities, municipal authorities and environment agencies. The World Bank, Washington, DC: p. 103.

Gogu, R.C. & Dassargues, A. 2000. Current trends and future challenges in groundwater vulnerability assessment using overlay and index methods. *Environmental Geology* 39(6): 549–559.

Hannapel, S. & Voight, H.J. 1999. Vulnerability maps as a tool for groundwater protection – case studies from Eastern Germany. In: *Hydrogeology and land use management* (M. Fendekova & M. Fendek eds.), Proc. XXIX Congress IAH Bratislava: 59–64,

Holting, B., Haertle, T., Hohberger, K.H., Nachtigal, K.H., Villinger, E., Weinzierll, W. & Wrobel, J.P. 1995. Konzept zur Ermittlung der Schutzfunktion der Grundwasseruberdeckung. *Geol. Jahrbuch*, Reihe C, Heft 63, Hannover.

Howard, K.W.F. 1985. Denitrification in a major limestone aquifer. *Jour. of Hydrology* 76: 265–280.

Identification... 2003. Identification of water bodies. Horizontal guidance document on the application of the term "water body" in the context of the Water Framework Directive. *Common Implementation Strategy for the WFD (2000/60/EC)*. Brussels.

Juergens-Gschwind, S. 1989. Ground water nitrate in other developed countries (Europe) – relationships to land use patterns. In: *Nitrogen management and groundwater protection; Ser. Developments in agricultural and managed-forest ecology*, 21 (Follett ed.), Amsterdam: Elsevier: 75–125.

Kazimierski, B. & Macioszczyk, A. 2003. Groundwater monitoring under new law regulations. In: *Wspolczesne Problemy Hydrogeologii*, v. XI (Piekarek-Jankowska & Jaworska-Szulc eds.), Gdansk Univ. of Technology, Gdansk (in Polish): 399–410.

Kleczkowski, A.S. (ed). 1990. The map of the Critical Protection Areas (CPA) of the Major Groundwater Basins (MGWB) in Poland, 1:500,000. Central Research Program "Environmental Management and Protection", AGH – Univ. of Science and Technology, Krakow: p. 44.

Knisel, W.G. & Davis, F.M. 1999. GLEAMS: Groundwater loading effects of agricultural management systems, ver. 3.0. User manual., USDA-ARS-SEWERL, Trifton, Georgia: p. 182.

Nawrocki, P. & Madgwick, J. 1999. The status of wetlands in Poland. WWF European Freshwater Program.

Paczynski, B., Plochniewski, Z. & Sadurski, A. 1999. Hydrogeological map of Poland 1:50,000 – new stage of Polish hydrogeological cartography. *Bull. Polish Geological Institute* 388 (in Polish): 191–210.

Prazak, J., Witczak, S. & Zurek, A. 2001. The problems of groundwater disposable resources assessment in river catchments where streamflow rate is controlled by groundwater baseflow. In: *Wspolczesne Problemy Hydrogeologii*, v. X T. (Bochenska & S. Stasko eds, Wroclaw, Publishing Office Sudety (in Polish): 235–243.

Refsgaard, J.C., Thorsen, M., Jensen, J.B., Kleeschulte, S. & Hansen, S. 1999. Large scale modeling of groundwater contamination from nitrate leaching. *Journal of Hydrology* 221: 117–140.

Robins, N., Adams, B., Foster, S. & Palmer, R. 1994. Groundwater vulnerability mapping: the British perspective. *Hydrogeologie* 3: 35–42.

Stockmarr, J. 2001. Grudvandsovervagning 2001. GEUS, Copenhagen.

Vrba, J. & Zaporozec, A. (eds.) 1994. Guidebook on mapping groundwater vulnerability; IAH Intern. Contribution to Hydrogeology, v.16, Hannover: Heise Verlag.

Winter, T.C., Harvey, J.W., Franke, O.L. & Alley, W.M. 1998. Ground water and surface water: a single resource. *US Geological Survey Circular*, 1139, Denver, Colorado.

Witczak, S. & Zurek, A. 2000. Evaluation wellhead protection area using porosity different types; *Proc. XIII PZITS Symp. "Groundwater municipal supply and management problems"*, Czestochowa (in Polish): 109–115.

Witczak, S., Duda, R., Zurek, A. & Szklarczyk, T. 2003. Groundwater flow model of different hydro-geological systems. In: *Wspolczesne Problemy Hydrogeologii.* v. XI (Piekarek-Jankowska & Jaworska-Szulc eds.), Gdansk Univ. of Technology, Gdansk (in Polish): 481–489.

Witkowski, A.J., Kowalczyk, A., Rubin, K. & Różkowski, A. 2002. The concept of groundwater vul-nerability maps on example Silesian Trias fissured-karst aquifers. In: *Groundwater quality and vulnerability, Prace Wydziału Nauk o Ziemi*, 22, (Rubin, Rubin & Witkowski eds.), Univ. of Silesia, Sosnowiec (in Polish).

Worrall, F. & Burt, T.P. 1999. The impact of the land – use change on water quality at the catchment scale: the use of export coefficient and structural models. *Journal of Hydrology* 221: 75–90.

Zurek A. 2002 – Nitrates in groundwater – an overview. *Bull. Polish Geological Institute* 400 (in Polish): 115–141.

Zurek, A., Witczak, S., Duda, R. 2002. Vulnerability assessment in fissured aquifers. In: *Groundwater quality and vulnerability. Prace Wydziału Nauk o Ziemi*, 22, (Rubin, Rubin & Witkowski eds.), Univ. of Silesia, Sosnowiec (in Polish): 241–254.

Case studies

Part I.   Porous aquifers

CHAPTER 5

Contamination of coastal aquifers from intense anthropic activity in southwestern Sicily, Italy

A. Aureli,[1] A. Contino,[1] G. Cusimano,[1,3] M. Di Pasquale,[4] S. Hauser,[2,3]
G. Musumeci,[4] A. Pisciotta,[3] M.C. Provenzano[3] & L. Gatto[1]

[1] *Dipartimento di Geologia e Geodesia, Università di Palermo, Italy.*
[2] *Dipartimento di Chimica e Fisica della Terra (C.F.T.A.), Università di Palermo, Italy.*
[3] *Istituto Nazionale di Geofisica e Vulcanologia (Sezione di Palermo), Italy.*
[4] *Unità Operativa 4.17 del G.N.D.C.I (C.N.R), Palermo, Italy.*

ABSTRACT: This study was conducted in western Sicily, in the Marsala and Castelvetrano-Campobello di Mazara coastal plain area. The economy of the area is mainly based on intense farming of specific agricultural crops (citrus, grapes, olives and others in greenhouses). These activities require a substantial and growing amount of water drawn, essentially, from local aquifers. In summers, water demand increases due to high population density. In order to satisfy this demand, numerous wells were drilled throughout the area, in some cases with a density of 10 wells/km^2. The uncontrolled groundwater exploitation in the last 10 years has resulted in a drastic decrease in water well levels, consequently allowing extensive seawater intrusion. The main aquifer is a sandy-calcarenitic, Plio-Pleistocenic complex, interbedded with clayey-marly layers in the lower layers. Multi-layered aquifers are present with the upper layers unconfined. Related studies are also being conducted in this area including monitoring of over 100 wells to better define potential groundwater resources.

More specifically, the authors of this study, who are part of the U.O. 4.17 of the Gruppo Nazionale Difesa dalle Catastrofi Idrogeologiche (G.N.D.C.I) (National Defense Group for Hydrogeologic Catastrophes) of the Consiglio Nazionale delle Ricerche (C.N.R.) (National Research Council), are completing a "Pollution Vulnerability Map" of all aquifers in south-western Sicily (1:50.000 scale) using the zoning method for homogeneous areas (G.N.D.C.I.-C.N.R. method). This map is a valid instrument for all authorities governing the management and protection of groundwater resources in the area.

1 GEOGRAPHICAL SETTING

This study was conducted on the coastal plain area of Marsala and Castelvetrano-Campobello di Mazara, (SW Sicily, ~900 km^2), between the Fiumara di Marsala riverbed and the Delia and Modione rivers. The cities within this area are: Marsala, Mazara del Vallo, Castelvetrano and Campobello di Mazara (Figure 1).

The Marsala and Castelvetrano-Campobello di Mazara plain morphology is relatively flat with less than 20% inclination and elevations that vary between a few meters a.s.l. along the coastline to approximately 250 m, in the adjacent hills inland. A distinct series of marine terraces (D'Angelo & Vernuccio 1996) up to 170 m a.s.l. can be easily identified

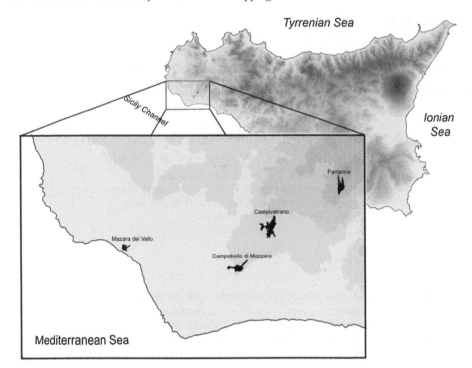

Figure 1.   The location of the study area in Southwest Sicily.

by the presence of a series of morphologic steps, leveled at the top by erosion, previously covered by paleosoils.

The Castelvetrano-Campobello di Mazara coast has large predominantly sandy beaches with sand dunes that have formed parallel to the coastline. Both the Marsala and Castelvetrano-Campobello di Mazara coast contain different wetlands, today partially or totally dried. These are locally known as *"margi"* or *"gorghi"*. Some of these areas have recently been declared protected areas or natural reserves for their environmental value (S.C.I. – Site of European Community Interest). Another area of particular scientific value is the "Preola Lake" and "Gorghi tondi" karst area manifesting evidence of chemical dissolution phenomena of underlying gypsum below the Pleistocene covering.

2   GEOLOGICAL FRAMEWORK

The studied area contains Pleistocene marine deposits (Calcareniti di Marsala *Auct.*) of sand and bioclastic calcareous gravel changing laterally and vertically, to calcarenites and calcirudites (Ruggieri et al. 1977; D'Angelo & Vernuccio 1992, 1994), that unconformity cover the marly-arenaceous Valle del Belice Formation (Ruggieri & Torre 1973; Vitale 1990), made up of a terrigeneous Plio-Pleistocenic sequence (sandstones and calcarenites with clay interbedded). This is related to the outcropping mountain chain areas that provided clastic fragments found in the surrounding basin areas ("Fossa di Castelvetrano"). Further deeper, underlying the Pliocene inf. (Trubi) marly calcilutites, evidence of the Messinian evaporite succession (evaporitic limestone and gypsum), laying unconformity on conglomeratic and/or sandy and marly clay deposits of the Cozzo Terravecchia Formation *Auct.* (Flores 1959; Schmidt di Friedberg 1962). The Pleistocene marine deposits (sands, gravels and calcarenites, max. 10 m thick) terraces are found in no specific order (D'Angelo & Vernuccio 1996; D'Angelo et al. 2001).

Palustrines, dunes and eluvium colluvial deposits can be found along the coast and ter-raced alluvium are found near the main rivers. In the NE sector of the Marsala Plain, Messinian and medium-Pliocene tectonics created several folds (NE-SW fold axis direc-tion). Meanwhile, in the Castelvetrano-Campobello di Mazara Plain, the same tectonic events created a synclinal structure (NE-SW axis direction) plunging SW. The following tectonic phases, mainly extensional, involved the Pleistocene calcarenite to post-Tyrrhenian deposits.

3   HYDROGEOLOGICAL FRAMEWORK

Seven hydrogeological complexes were identified, each playing a distinct role in ground-water circulation depending on geometrical relationship and permeability of the complex:

- Clayey-marly complex – includes the Cozzo Terravecchia Formation, with a very low level of permeability due to porosity. This makes up the regional impermeable base-ment of the underlying aquifers.
- Calcareous complex (*Calcari di base* and *Calcari a Porites* of Messinian evaporitic succession), with medium-low to high permeability due to fracturing and karst phenomena.
- Gypsum complex, including gypsum, gypsolutite and gypsoarenite (evaporitic succes-sion) with low to medium-high permeability due to fracturing and karst phenomena. The groundwater flow generally follows fractures and karstic conduits, sometimes feeding springs with very low discharge, manifestations along the contact with under-lying clay of the Cozzo Terravecchia Formation.
- Marly complex, including the marly calcareous terrains and Pliocenic clayey-marly ter-rains (*Trubi*) with varied permeability, from very low to medium-high, due to fracturing. This is the impermeable layer under the marly-arenaceous Valle del Belice Formation.
- Marly-calcarenitic complex includes the predominantly arenaceous-calcarenitic lower part of the marly-arenaceous Valle del Belice Formation. It has a medium-high permea-bility first due to porosity and secondly due to fracturing.
- Argillaceous complex includes the clay of the early Pleistocene and the clayey portion of the marly-arenaceous Valle del Belice Formation. That finds itself interbedded, sometimes, through the calcarenitic complex. It has a very low permeability due to porosity and consequently an insignificant impermeability role.
- Calcarenitic complex includes the arenaceous-conglomeratic terrains and sandy parts of the marine terraces, the *Calcarenite di Marsala* and a portion of the upper calcarenites from the marly-arenaceous Valle del Belice Formation. It has a medium-high per-meability, primarily due to porosity and subsequently due to fracturing. Some authors (Dall'Aglio & Tedesco 1968) have studied the Staglio and Gaggera springs (Modione river), located at the contact between the calcarenitic complex and the Plio-Pleistocenic

clays that, up till 1966, had at over 50 l/s discharge. Presently, these springs have completely dried up due to piezometric level lowering from uncontrolled pumping wells.

• Alluvium complex plays a secondary hydrogeological role. It is a multi-layered aquifer that is in communication with the calcarenitic aquifer.

• A mainly impermeable complex of recent terrain deposits (palustrines, black soils, etc.) does not play any hydrogeological function due to its limited extension.

Two study areas are considered to be of notable hydrogeological importance:

Marsala plain hydrogeological unit, located from the final part of the *Fiumara di Marsala,* riverbed – north and the Delia River – south. The plain between Marsala and Mazara del Vallo contains a multi-layered aquifer. The layers are conditioned by varied permeability of different calcarenitic levels with intercalation of low permeable silty-clays. The groundwater in the different layers come into contact with each other in areas where clayey levels are thinned out or interrupted. The aquifer has, as basement, the clays and/or marls of the Cozzo Terravecchia Formation. The calcarenitic aquifer thickness varies from a few meters to approximately 70 m. In the last 50 years, the depth to water in Mazara del Vallo area have increased, today it is presently about 22 m. This phenomenon is also confirmed by the dewatering of some significant springs (20 l/s, 1933).

Castelvetrano-Campobello di Mazara plain hydrogeological unit laterally delimitated by Tortonian clayey-sandy deposits outcropping west along the Delia River and east of the early Pleistocene clay and the clayey lithofacies of the marly-arenaceous Valle del Belice Formation outcropping along the Modione River. The stratigraphic analysis of drinking water wells (Staglio and Bresciana) revealed the presence of a multi-layered aquifer hosted in the Plio-Pleistocenic calcarenitic clayey sequence whose substratum coincides with the Tortonian clayey-marly hydrogeological complex.

The multi-layered aquifer is made up of:

– a shallow unconfined aquifer located in the upper calcarenitic portion with reduced thickness and a variable saturated zone (1 m to ~10–20 m);

– a deep semi-confined aquifer contained in the calcarentic-marly hydrogeolocial complex. It has elevated thickness, where the more massive portion of the complex (150 m) shows a very high transmissivity average (~5 $\times$ 10^{-2} m^2/s).

South of the Campobello di Mazara urban area, the multi-layered aquifer becomes an unconfined aquifer due to level thinning and the eventual complete absence of interbedded aquitards.

4   CONCEPTUAL HYDROLOGICAL MODELS OF COASTAL PLAINS

The hydrogeological model proposed for SW Sicily is complex and presents interconnection to seawater intrusion due to brought on by excessive groundwater pumping.

4.1   *Marsala Plain*

The piezometric lines of the calcarenitic aquifer, intensely exploited, are shown in Figure 2. That aquifer has discontinuous layers of different permeability that cause, locally, the semi-confinement of deep groundwater (survey November 1999, Calvi et al. 2001).

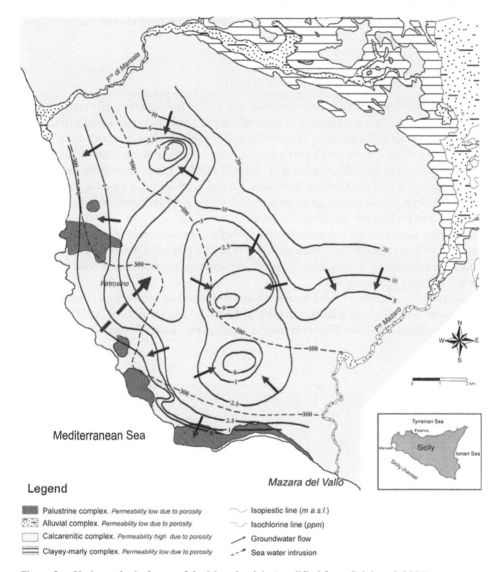

Mediterranean Sea

Mazara del Vallo

Legend

▮ Palustrine complex. *Permeability low due to porosity*
▦ Alluvial complex. *Permeability low due to porosity*
☐ Calcarenitic complex. *Permeability high due to porosity*
▤ Clayey-marly complex. *Permeability low due to porosity*

〰 Isopiestic line (*m a.s.l.*)
〰 Isochlorine line (*ppm*)
↗ Groundwater flow
⤍ Sea water intrusion

Figure 2. Hydrogeological map of the Marsala plain (modified from Calvi et al. 2001).

This area has 25 exploited wells that are managed by the Marsala Municipality and the *"Ente Acquedotti Siciliani"*. Reconstruction of piezometric surface, trough 53 monitoring wells, has allowed to draw the groundwater divides and groundwater flow: one area where the groundwater flow direction goes towards the sea and another where the flow is modified and conditioned by pumping wells (south-west Petrosino). In fact, the amount of water pumped exceeds the amount of annual recharge. This causes a visible lowering of the piezometric level, and in some cases, cuts water yield by 50% in some wells, which are consequently periodically deepened. As a result, water resource impoverishment has led to the disappearance of some wetlands (*margi*) but above all, has led to seawater intrusion phenomena (Cl variable from 250 to 950 ppm), deteriorating groundwater quality (Calvi et al. 2001; Hauser et al. 2002).

4.2   *Castelvetrano-Campobello di Mazara Plain*

The piezometric lines related to deep semi-confined aquifer (calcarenitic marly complex) are shown in Figure 3. That is connected, in the southern plain area, with the unconfined aquifer (survey November 1999, Bonanno et al. 2000). In the northern plain area, groundwater discharge flows NE-SW towards Contrada Staglio where the Ente Acquedotti Siciliani well fields are located. In the southern plain area, groundwater flow follows two preferential directions, converging into the Contrada Bresciana, an area with intensive pumping (Bresciana wells). Evidence of submarine groundwater discharge, along the coastline between Torre Granitola and Tre Fontane, has also been found.

The main aquifer of the Castelvetrano-Campobello di Mazara plain is subject to intense pumping with an annual average recharge lower than 15% creating a deficit of $2.7 \times 10^6 \, \mathrm{m}^3$. This is the main cause of the gradual groundwater resource impoverishment and the visible lowering of the piezometric level.

The alarming situation of overexploitation was confirmed by comparison of piezometric levels of the semi-confined aquifer in autumn of 1999 (Bonanno et al. 2000) and those taken in 1981 (CasMez 1981), showing an average decrease in piezometric level of 20 m in the last 19 years.

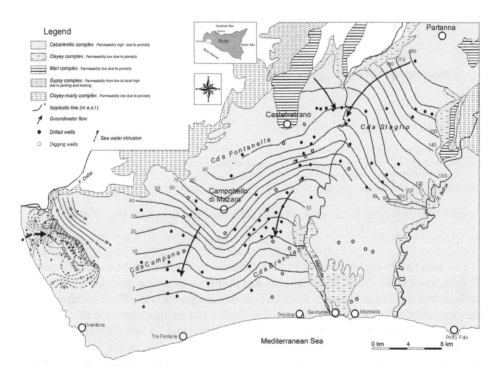

Figure 3.   Hydrogeological map of the Castelvetrano – Campobello di Mazara plain (modified from Bonanno et al. 2000).

In the Contrada Bresciana wells field, the groundwater lowering varies from 5 to 10 m. The high transmissivity value and deep groundwater recharge reduce the effects of excessive pumping changing the hydrodynamic conditions of the groundwaters. In the sector near to the coastal plain, piezometric level lowering varies between 5 to 20 m, probably due to pumping from the Contrada Bresciana wells field. (Bonanno et al. 2000). The excessive deep aquifer overexploitation is also confirmed by simulation model results calibrated in transitory regime, using the "*Processing Modflow*" (Ciabatti & Provenzano 2003) that shows that the hydrogeological system is not affected by vertical and lateral recharge variation.

5   QUALITATIVE CHARACTERISTICS AND POLLUTION VULNERABILITY

The study area is subject to intense agriculture activities (grapevines, olive orchards, citrus orchards etc.), which requires a considerable quantity of water that is mainly supplied by the pumping of deep aquifers.

The Figure 4 shows the spatial distribution of four vulnerability degrees computed by CNR-GNDCI method (zoning for homogeneous areas, Civita 1988). The highest degrees occur in the calcarenitic Complex, where the agricultural activity is very diffuse. This activity also employs massive use of organic and/or inorganic fertilizers, which infiltrate into the soil modifying groundwater quality. Moreover, for years open air and underground Pleistocene calcarenite quarries have operated in the two plain areas. Today, the abandoned

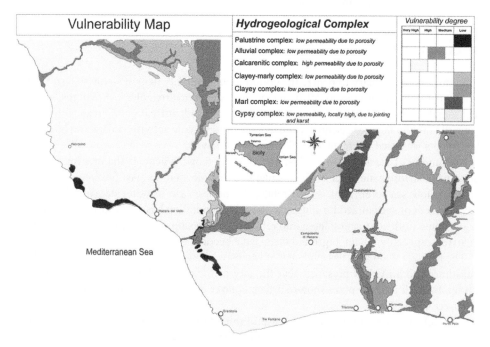

Figure 4.   Groundwater vulnerability map of the Marsala and Castelvetrano – Campobello di Mazara plains.

quarries are used as dumping grounds for solids and wastewaters, especially those near the city of Marsala; causing an increase in groundwater pollution.

This type of pollution phenomena has been documented in the shallow aquifer that is hydraulically interconnected with river water. Concentration values of nitrate, potassium and sulfate are over the drinking standard established levels by Italian Legislative Law (no. 152 of 11 May 1999). On the contrary, the deep groundwaters, rich in bicarbonates, do not register values over those allowed by law (Bonanno et al. 2000).

The Castelvetrano and Campobello di Mazara plain has high degree of intrinsic vulnerability. On the other hand, the deeper aquifer part has a medium-low degree. However, the discontinuous presence of clayey levels permits the infiltration of contaminants towards the deeper portion. Moreover, this area is also characterized by diffuse seawater intrusion due to coastal water wells overexploitation.

According to the Legislative Law (no. 152 of 11 May 1999), the aquifer water can be classified as:

– Deep aquifer class 2C
– Surface aquifer class 4C

where:

- C indicates "human impact effect on resource availability shown by modification of general indicators set by law";
- 2 indicates "reduced human impact sustainable for long periods of time with good hydrochemical characteristics";
- 4 indicates "considerable human impact with poor hydrochemical characteristics."

6   CONCLUSION

This study is preliminary to realize the pollution vulnerability map of southwestern Sicily. It includes detailed surveys to define the geological and hydrogeological features, as well as groundwater geochemical information, allowing the identification of areas with different vulnerability degree, essentially related to anthropogenic activity. A database of all collected data was used to create a G.I.S. using Arc View 3.2.

Monitoring of the area over the last 30 years has shown a general and progressive lowering of the hydrostatic level, with considerable drawdown in the last few years, reaching values below sea level. That is also related to the increase in water demand due to the development of specialized farming activities. This condition has also caused the dewatering of some wetlands turning them into refuse dumping sites. Furthermore, the abandoned quarries, especially the underground ones, have become disposal waste sites.

The Marsala Plain area has evidence of extensive seawater intrusion along the complete coastline with a marked presence near the city of Mazara del Vallo. The Castelvetrano-Campobello di Mazara plain shows surface aquifer quality deterioration due to massive fertilizer use.

In conclusion, this study will play a useful role in realizing the integrated vulnerability map for southwestern Sicily that is in progress. This map will be a valid tool for government management to utilize in decision-making processes for the protection of all underground water resources in the area.

REFERENCES

Bonanno, A., Ciabatti, P., Liguori, V., Provenzano, M.C. & Sortino, G., 2000. Studio idrogeologico ed idrogeochimico dell'acquifero multifalda della Piana di Castelvetrano e Campobello di Mazara (Sicilia occidentale). *Quaderni di Geologia Applicata* 7, 4: 45–59.

Ciabatti, P. & Provenzano, M.C. 2003. Simulazione del flusso idrico sotterraneo dell'acquifero multifalda della Piana di Castelvetrano e Campobello di Mazara (Sicilia sud-occidentale). *Quaderni di Geologia Applicata* 10, 3: 5–16.

Calvi, F., Frias Forcada, A., Pellerito, S., 2001. *Regime idrodinamico indotto nel sistema acquifero costiero tra Marsala e Ma zara del Vallo (TP)*. *Acque Sotterranee* 74: 15–21.

Cassa per il Mezzogiorno 1981. Ripartizione progetti idrici, divisione V – Schemi idrici della Sicilia. Indagini idrogeologiche per l'approvvigionamento idrico del sistema II nord-occidentale della Sicilia (acquiferi principali).

Civita, M. 1988. Le Carte di vulnerabilità degli acquiferi all'inquinamento. In: Proposta di Normativa per l'istituzione delle fasce di rispetto delle opere di captazione di acque sotterranee. Pubbl. n. 75 CNR-GNDCI, Ed. V. Francani, Coed. M. Civita, Geo-Graph Segrate (Milano): 45–55.

Dall'Aglio, M. & Tedesco, C. 1968. Studio geochimico ed idrogeologico di sorgenti della Sicilia. *Rivista Mineraria Siciliana*, 112–114: 27–66.

D'Angelo, U. & Vernuccio, S. 1992. Carta Geologica del Foglio 617 "Marsala" (scala 1:50.000) – Dipartimento di Geologia e Geodesia dell'Università degli Studi di Palermo.

D'Angelo, U. & Vernuccio, S. 1994. Note illustrative della Carta Geologica del Foglio 617 "Marsala" (scala 1:50.000). *Boll. Soc. Geol. It.*, CXIII: 55–67.

D'Angelo, U. & Vernuccio, S. 1996. I terrazzi marini quaternari dell'estremità occidentale della Sicilia. *Mem. Soc. Geol. It.*, LI: 585–594.

D'Angelo, U., Parrino, G. & Vernuccio, S. 2001. Il Quaternario della fascia costiera compresa fra la punta Granitola e Porto Palo (Sicilia sud occidentale). *Naturalista Siciliano*, s. IV, XXV, n. 3–4: 333–344.

Flores, G. 1959. Evidence of slump phenomena (olistostromes) in areas of hydrocarbons exploration in Sicily. In: Proceedings of Fifth World Petroleum Congress., Sect. 11, Paper 13, New York: 259–275.

Hauser, S., Cusimano, G. & Vassallo, M. 2002. Idrogeochimiai di ambienti umidi costieri: Mazara del Vallo, Trapani. GEAM 107, 4: 71–76.

Schmidt di Friedberg, P. 1962. Introduction à la géologie pétrolière de la Sicile. *Rev. Inst. Franc. Pétr.*, 17: 653–688.

Ruggieri, G. & Torre, G. 1973. Geologia delle zone investite dal terremoto del Belice. 1) La Tavoletta di Gibellina. *Rivista Mineraria Siciliana*, 132–139: 127–187.

Ruggieri, G., Unti. A. Unti. M. & Moroni. A. 1977. La calcarenite di Marsala (Pleistocene inferiore) e i terreni contermini. *Boll. Soc. Geol. It.*, 94: 1623–1657.

Vitale, F.P. 1990. Studi sulla Valle del Medio Belice (Sicilia centro occidentale) – L'avanfossa plio-pleistocenica nel quadro dell'evoluzione paleotettonica dell'area. Tesi di dottorato.

CHAPTER 6

Vulnerability assessment of a shallow aquifer situated in Danube's Plain (Oltenia-region, Romania) using different overlay and index methods

I.J. Dragoi[1] & R. Popa[2]
[1]*Geographic Institute of the Romanian Academy, Bucharest, Romania*
[2]*Faculty of Geology and Geophysics, University of Bucharest, Romania*

ABSTRACT: For assessing intrinsic groundwater vulnerability of a shallow aquifer situated in the Romanian part of the Danube river floodplain three different methods have been used. Taking into consideration the morphology of the study area, characterized by sand dunes and regarding the accuracy of available data the best method to assess of the intrinsic vulnerability has been selected.

1 INTRODUCTION

The lack of or difficult access to digital data concerning Romania is one of the issues which makes vulnerability assessment using overlay and index methods a very hard and time consuming operation. While in other countries, in some cases, data collecting is reduced to looking on the internet, in Romania the first step of a vulnerability study means in fact digitizing maps and creating the necessary input parameter databases for the different assessment methods. So, from the beginning, a vulnerability study using GIS is expensive, and in most cases vulnerability maps are still obtained in the old fashion "manual" way. The purposes of the following study were:

- To emphasize the advantages of GIS techniques in groundwater vulnerability assessment;
- To present intrinsic vulnerability maps obtained using different methods (DRASTIC, GOD and AVI) and discuss the results;
- To determine the best method to use for vulnerability assessment of shallow aquifers situated in the floodplains of different representative rivers of Romania.

2 DESCRIPTION OF THE STUDY AREA

The study area is situated in the southern part of Romania in a region called Oltenia. This region is bounded on the north by the South Carpathian chain, on the west and south by the Danube River and on the east by the Olt River.

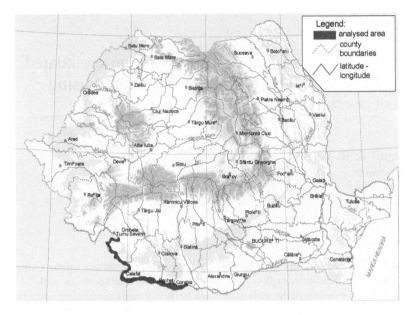

Figure 1.   Map showing the location of the study area.

The most developed morphological structure of Oltenia region is a flat plain, which is part of the Romanian plain. In this plain some the most important rivers of Romania (such as the Danube, Olt and Jiu rivers) have created large floodplains and their terraces can be well observed (1969, Geografia Vaii Dunarii Romanesti).

In the alluvial deposits of these three rivers (mostly of Quaternary age), large shallow aquifers are developed. For this study we chose to assess the intrinsic vulnerability of a small part of the regional shallow aquifer connected to the Danube River, the part situated in the floodplain and the area of inferior terraces of this river (Figure 1). The different conditions of formation of the Danube valley are reflected in the actual configuration of the investigated area that is characterized by the existence of a flat floodplain, by terraces that can be more or less observed and by sand dunes (which can be observed on the maps in fig. 6 -with a rating of 0.6- and in fig. 12 -with a rating of 7).

The Danube River floodplain has more developed sectors such as downstream from Calafat town and less developed ones as in the Cetate village area (Figure 2). The particular aspect of the Danube river floodplain is conferred by the presence of strong aeolian activity, which is so intense that a large part of the floodplain is not now in fact under the influence of the Danube River, even in case of large floods.

The terraces are well developed at Crivina (in the north-western part of the area), Ciuperceni (central part), Corabia and Dabuleni (eastern part). The absolute level (above level 0 of the Black Sea) of the Corabia terrace is situated between 40 and 60 m and is covered by sand dunes. The Ciuperceni terrace has absolute levels of between 35 and 45 m and due to intense aeolian activity, is covered almost all over by sand dunes, as we can see at Desa village (Figure 3).

The permanent lakes in the area are isolated or interconnected (Ciuperceni, Rast and Bistret lakes) and sometimes the groundwater level rises higher then the surface level, creating temporary lakes.

Figure 2.   The Danube floodplain upstream of Calafat town (Cetate village area).

Figure 3.   Sand dune at Desa.

Vegetation is poor, with small and isolated forests that have no influence on aquifer recharge. From a land use point of view, the greater part of the area is used for agricultural purposes but since 1980 the intensive use of fertilizers has ceased so there is no problem with nitrate pollution. Some villages and two towns represent the few populated localities in the area. The villages have small populations as have Calafat and Corabia towns. There is no important industrial activity in the area.

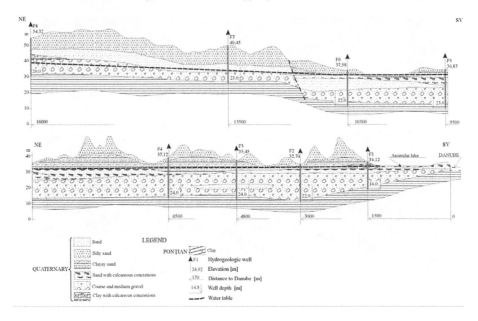

Figure 4.   Cross-section along the Ciuperceni monitoring well line.

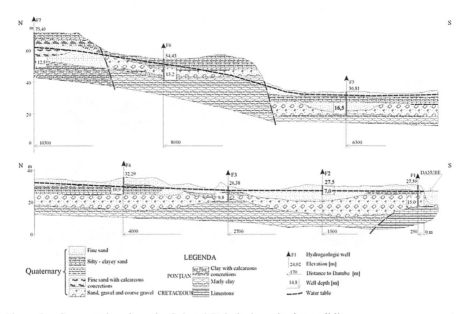

Figure 5.   Cross-section along the Calarasi-Dabuleni monitoring well line.

2.1   *Geology and hydrogeology*

From a regional geotectonic point of view, the study area belongs to the Moesian Platform. Of its four sedimentation cycles, only the last one (Badenian-Pleistocene) has importance

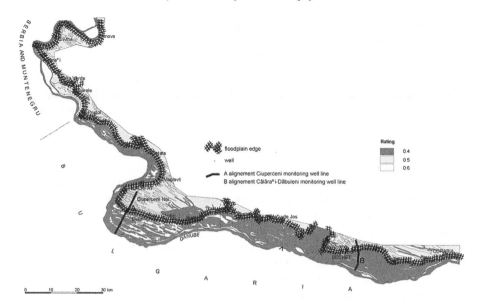

Figure 6. Assessment of overlying lithology (after ratings assignment) and positions of monitoring wells.

for the studied aquifer. The Pontian layers of this cycle represent the clayey bedrock of the aquifer (Figures 4 and 5; cross-section position Figure 6).

The studied aquifer is developed in Quaternary deposits (the part situated in the flood-plain and lower terraces is in Holocene deposits and the part situated in the upper terraces is in Pleistocene deposits) of the following types: sands, coarse sands, clayey sands and gravels. The thickness of the aquifer ranges from 0.5 m to 26 m and the hydraulic gradient has a small slope (0–0.001) toward the Danube. The shallow aquifer in the Oltenia Plain has a greater extension than the studied area, which is only the part situated Danube flood-plain and lower terraces.

3   DATA COLLECTING AND MANAGEMENT

Some of the necessary spatial data were digitized from different paper maps georeferenced to a local projection system, Stereo70. The paper maps used were:

– soils map 1:200,000 – published by the Pedologic Institute of Romania in 1986
– hydrogeological map 1:100,000 – published by Geological and Geophysical Institute of Romania in 1978
– hydro-geographical map of Oltenia Plaine 1:300,000 – published by the Institute of Geography of Romania in 1969 (contains information about surface water and the shallow aquifers of the Danube valley).

The attributes database contains the following information from 30 hydrogeological wells (Figure 6) situated in that area:

– an average piezometric level based on the annual mean level measured between 1965 and 1995
– well elevation

– hydraulic conductivity values from pumping test executed in each well
– the aquifer lithology and thickness
– lithology and thickness of each sedimentary unit above the aquifer.

Other spatial data (such as depth to water table, obtained as difference between well elevation and piezometric level and the aquifer conductivity), where interpolated from point data. Some of the input parameters (like slope) are considered constant all over the area.

4   INTRINSIC VULNERABILITY ASSESSMENT

The intrinsic vulnerability assessment methods used are GOD, AVI and DRASTIC. Intrinsic vulnerability is considered according to Vrba and Zaporozec (1994) as "*a function of hydrogeological factors – the characteristics of an aquifer and the overlaying soil and geological materials*".

GOD rating system (Foster 1987) is an empirical method for quick assessment of vulnerability using three parameters: Groundwater occurrence, Overlaying lithology and Depth to groundwater. To each category of these parameters a rating is assigned. The vulnerability index is calculated with the following formula:

> *GOD vulnerability index = Rating for Groundwater occurrence ×*
> *Rating for Overlaying lithology (only in case of unconfined aquifers) ×*
> *Rating for Depth to water*

The studied aquifer is unconfined so the rating for groundwater occurrence is the same for the whole are. The overlaying lithology was assessed with the help of the hydro-geographical map. The assigned rating values and the distribution of this parameter is shown in Figure 6.

The depth to groundwater was assessed using point data from monitoring wells and using an inverse distance method for interpolation. The resulting distribution, after assigning the rating values proposed for that method, is shown in Figure 7.

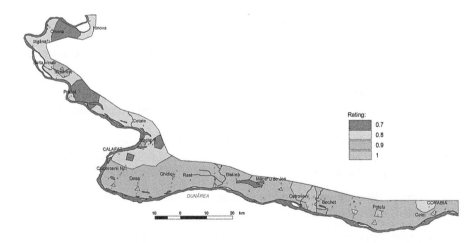

Figure 7.   Depth to groundwater assessment (after ratings assignment).

The resulted intrinsic vulnerability map (Figure 8) indicates higher vulnerability in the area where the sand dunes are well developed and clearly indicates the position of the dunes.

4.1 *AVI rating system (van Stempvoort et al. 1993)*

In this method two parameters are considered: the thickness of each sedimentary unit above the uppermost aquifer (d) and the estimated hydraulic conductivity of each of these layers (k).

The hydraulic resistance (c), calculated by the following equation, is related to a qualitative Aquifer Vulnerability Index by a relationship table.

$$c = \sum_{i=1}^{n} \frac{d_i}{k_i}$$

where:

– n is the number of sedimentary units above the aquifer
– d the thickness of each sedimentary unit above the uppermost aquifer
– k estimated hydraulic conductivity of each sedimentary unit.

To assess the intrinsic vulnerability using the AVI method some authors (Gogu & Dassargues 2000, Wei 1996) suggest calculation of the hydraulic resistance for each borehole and creation of an iso-resistance map. We have used the same methodology and the resulting vulnerability map is shown in Figure 9 together with the resistance – vulnerability class relationship table (legend).

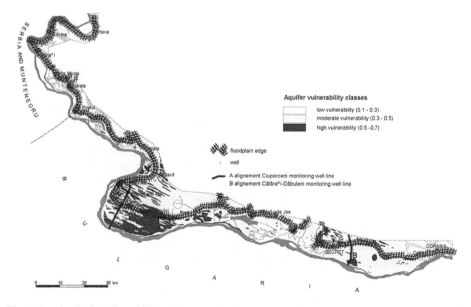

Figure 8.   Intrinsic vulnerability map according to GOD method.

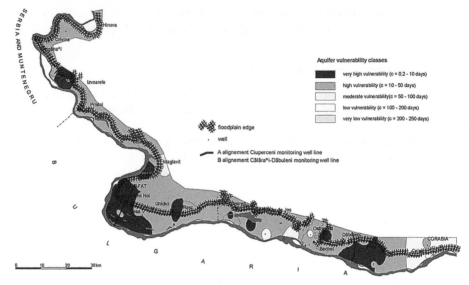

Figure 9.   Intrinsic vulnerability map according to the AVI method.

This also indicates also higher vulnerability for the sand dunes area situated near Ciuperceni and Desa villages (in the central part of the studied area), a fact that is in accordance with the hydrogeological conditions in that area but the position of the sand dunes themselves is not well observed. At the same time, assessment using this method is more complicated.

4.2   *DRASTIC point count system model (Aller et al. 1987)*

In this method a vulnerability index for each mapping unit is calculated using seven different parameters, using the following formula:

DRASTIC Index = Dr × Dw + Rr × Rw + Ar × Aw + Sr × Sw + Tr × Tw + Ir × Iw + Cr × Cw

Where: **Dr, Dw** = Rating and weight for Depth to Groundwater; **Rr, Rw** = Rating and weight for net Recharge; **Ar, Aw** = Rating and weight for Aquifer media; **Sr, Sw** = Rating and weight for Soil media; **Tr, Tw** = Rating and weight for Topography/slope; **Ir, Iw** = Rating and weight for Impact of vadose zone; **Cr, Cw** = Rating and weight for hydraulic Conductivity.

Each parameter is assigned the same weight all over the area but different ratings, according to the hydrological, geological and hydrogeological conditions in the area. In our study some parameters (like slope and aquifer media) had the same rating all over the area, due to their small variability (see Table 1).

The assigned rating for the parameters with high variability and their distribution is shown in Figures 10 to 13 and the relationship parameter-rating in Table 2.

Table 1.   Weights and ratings used for DRASTIC

Parameter	Weight	Rating
Depth to groundwater	5	5–10
Net recharge	4	7–10
Aquifer media	3	8
Soil media	2	1–10
Topography/slope	1	10
Impact of vadose zone	5	4–7
Hydraulic conductivity	3	1–10

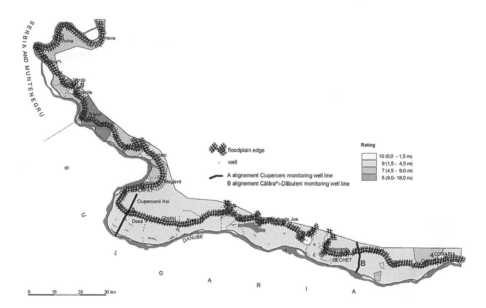

Figure 10.   Groundwater depth (after rating assignment).

The vulnerability classes were defined according to the range of the obtained vulnerability index, taking into account a classification made by Navulur et al. (1997), but even the classification proposed by Lobo Ferreira & Oliveira (2003) will show the same vulnerability in the area of interest. The intrinsic vulnerability map obtained (Figure 14) also indicates higher vulnerability in the sand dune area but the exact location of the sand dunes is not well defined.

At the same time, a more precise assessment of the intrinsic vulnerability using this method needs more accurate data for the parameters that have maximum weights (like depth to groundwater). In our study the assessment of this factor was made using point data and interpolation, as we had no DEM and water level model with which to calculate more accurately the variation of the depth to water (Savulescu et al. 2000).

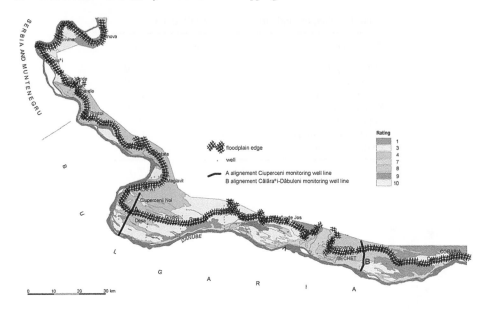

Figure 11.   Distribution of soil media (after rating assignment).

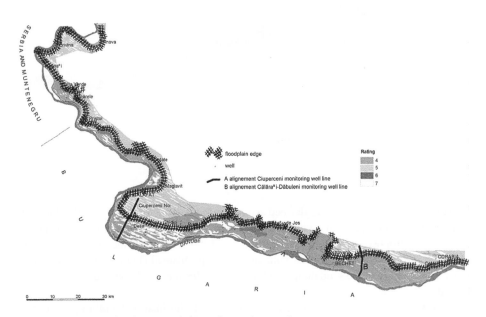

Figure 12.   Impact of vadose zone (after rating assignment).

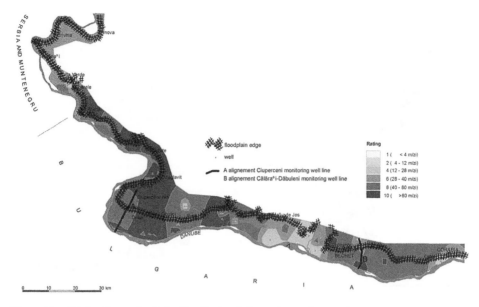

Figure 13.   Hydraulic conductivity distribution (after rating assignment).

Table 2.   Relationship parameter and rating for DRASTIC method

Parameter	Class	Rating
Depth to groundwater	0–1.5 m	10
	1.5–4.5 m	9
	4.5–9 m	7
	9–18 m	5
Soil media	Brown soil	1
	Chernozems	3
	Red brown wet soil	4
	Alluvial soil	7
	Regosol/sandy	8
	Mud soil	9
Impact of vadose zone	Aeolian sand	7
	Alluvial sand	6
	Silty sand	5
	Clayey sand	4
Hydraulic conductivity	<4 m/day	1
	4–12 m/day	2
	12–28 m/day	4
	28–40 m/day	6
	40–80 m/day	8
	>80 m/day	10

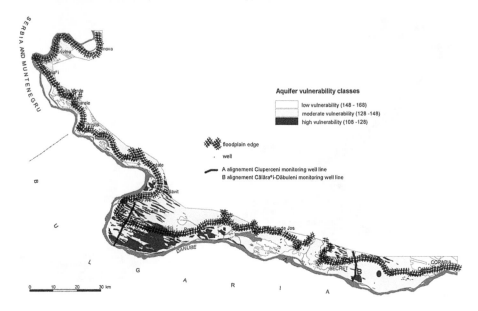

Figure 14.   Intrinsic vulnerability map according to the DRASTIC method.

5   CONCLUSIONS

The presence of such well-developed geomorphological structures as sand dunes (aeolian sand deposits) in a floodplain (where the deposits are mostly alluvial sand or coarse sand) should be observed in an intrinsic vulnerability map.

On all vulnerability maps regardless of the methods used, the area of the sand dunes is characterized by higher vulnerability.

Using the AVI and DRASTIC methods, more data are needed, the assessment is more time consuming and in the end the results are similar to those obtained with a simpler method like GOD.

At the same time, the presence of the sand dunes can be even better observed in the vulnerability assessment using the GOD method.

For the study area, where no special hydrological events and no intensive human activity is present, the vulnerability assessment does not need so many parameters.

In this case, for a quick assessment of the intrinsic vulnerability to pollution of aquifers situated in areas such as the studied one, we recommend the GOD method.

But in any case, starting from the moment that all the necessary digital data are available, groundwater vulnerability assessment can be made easily with the help of GIS techniques, no matter what method it is used.

REFERENCES

Aller, L., Bennett, T., Lehr, J.H., Petty, R.J. & Hackett, G. 1987. DRASTIC: a standardized system for evaluating groundwater pollution potential using hydrogeologic settings. US-EPA Report 600/2-87-035, Washington DC, USA.

Gogu, R.C. & Dassargues, A. 2000. Current trends and future challenges in groundwater vulnerability assessment using overlay and index methods, *Environmental Geology* 39 (6):549–555, Springer Verlag.

Foster, S.S.D. 1987. Fundamental concepts in aquifer vulnerability, pollution risk and protection strategy. *Proceedings International Conference VSGP*. Noordwijk – The Netherlands.

Lobo Ferreira, J.P. & Oliveira, M. 2003. On the experience of Groundwater Vulnerability Assessment in Portugal, Aquifer Vulnerability and Risk International Workshop AVR03, Salamanca Gto. Mexico.

Navulur, K.C.S. & Engel, B. 1997. Predicting Spatial Distributions of Vulnerability of Indiana State Aquifer System to Nitrate Leaching using GIS, http://www.ncgia.ucsb.edu/conf/ SANTA_FE_CD-ROM/sf_papers/navulur_kumar/my_paper.html.

Savulescu, C., Sârghiuta, R., Abdulamit, A., Bugnariu, T., Turcu, L. & Barbu, C. 2000. Fundamente GIS, Editura *H*G*A* Bucuresti.

van Stempvoort, D., Everet, L. & Wassenaar, L. 1993. Aquifer vulnerability index: a GIS compatible method for groundwater vulnerabilty mapping. *Can. Wat. Res. J.* 18: 25–37.

Vrba, J. & Zaporozec, A. 1994. Guidebook on Mapping Groundwater Vulnerability, International Association of Hydrogeologists (International Contribution to Hydrogeology 16), Heinz Heise Verlag, Hannover.

Wei, M. 1996. Evaluating AVI and DRASTIC for Assessing Groundwater Pollution Potential in the Fraser Valley, CWRA 51st Annual Conference Proceedings, Victoria, BC.

CHAPTER 7

Application of a groundwater contamination index to assessment of confined aquifer vulnerability

K. Dragon
Department of Hydrogeology and Water Protection, Institute of Geology, Adam Mickiewicz University, Poznan, Poland

ABSTRACT: A simple approach to the assessment and visualisation of groundwater contamination in a confined aquifer (the Wielkopolska Buried Valley aquifer, Poland) uses a groundwater
contamination index. The areas characterised by the most intense pollution groundwater due to
anthropogenic contamination were investigated. These areas were found to be the most vulnerable
parts of the Wielkopolska Buried Valley aquifer. The results of the research are consistent with previous studies completed by means of a multivariate statistical technique (factor analysis) as well as
conventional techniques. The contamination index is a simple method of assessment and carto-
graphic visualisation of groundwater contamination.

1 INTRODUCTION

Investigation of the spatial variability of groundwater chemistry (particularly changes of
groundwater chemistry due to human activity) provides a valuable vulnerability assessment
technique. The intensity of anthropogenic groundwater pollution can be used to distinguish zones of different vulnerability (Limisiewicz 1998). The components, which naturally
occur in groundwater at low concentrations, are used as the indicators of groundwater pollution. The change in concentration of these components can be directly related to the
influence of anthropogenic contamination.

Cartographic presentation of the overall degree of groundwater contamination is usually
based on the distribution of contaminant concentrations. A large number of separate maps
(for each indicator of groundwater contamination) are required to obtain a general view of
groundwater contamination.

A simple method for assessment and visualisation of groundwater contamination based
on the contamination index has been proposed. The method was successfully tested for
shallow groundwater in Finland and Slovakia by Backman and others (1998).

In the present study a confined aquifer – Wielkopolska Buried Valley aquifer (WBV)
has been chosen in order to test and apply the contamination index. The hydrogeologic and
hydrochemical condition of this aquifer as well as the influence of human activity on
groundwater chemistry are well documented for this region. This enables a comparison of
the results of this work with other investigations using different methods (i.e. multivariate
statistical methods applied by Dragon (2002).

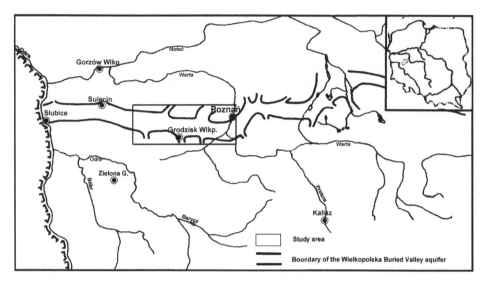

Figure 1.   Location scheme.

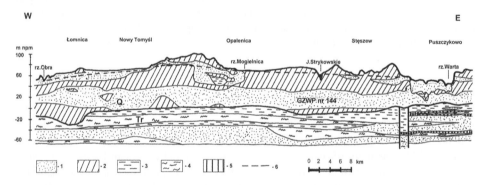

Figure 2.   Hydrogeological cross-section (after Dąbrowski 1990, simplified).
1 – sand and gravel; 2 – till; 3 – clay; 4 – silt; 5 – brown coal; 6 – groundwater level in the Wielkopolska Buried Valley aquifer; Q – Quaternary; Tr – Tertiary.

2   STUDY AREA

The WBV aquifer is one of the biggest aquifers in the middle of Great Poland. The study area within the WBV aquifer comprises about 1,000 km^2 (Figure 1).

The thickness of water bearing sediments (mainly fluvioglacial sands and gravels) ranges from 20 to 50 m (Figure 2). Tertiary clays hydraulically isolate the WBV aquifer from the underlying Tertiary aquifer. The confining layer is composed of glacial tills, which has a diverse thickness, ranging from about 20 m (in the regions of intertill aquifers and glacial troughs) to more than 50 m. However, the aquitard is absent in the Warta valley. The intertill aquifers occur mainly in the region of the Lwówek-Rakoniewice Rampart.

The main recharge area is located in the region of the Lwówek-Rakoniewice Rampart, from where groundwater flows westward to the Obra valley and eastward to the Warta valley. The principal source of recharge is percolation of rainwater through the glacial tills

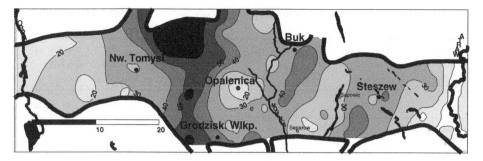

Figure 3.   Thickness of the Wielkopolska Buried Valley aquitard (in meters).

and upper intertill aquifers. Recharge by the inflow from the intertill aquifers located north of the WBV aquifer also occurs.

The aquifer is characterised by varied vulnerability. The most vulnerable areas occur in the regions where the thickness of the aquitard decreases to less than 30 m (Figure 3).

3   METHODS

The contamination index has been used in order to analyse groundwater contamination. This method has been carried out for the assessment and visualisation of the areas where the components of groundwater exceed the upper permissible limits and/or indicated values of potentially harmful elements, and where the concentration exceeds these limit values for shallow groundwater in Slovakia and Finland (Backman et al. 1998).

In present study the contamination index was used for assessment and mapping groundwater contamination degree of a confined aquifer (WBV aquifer). The first symptoms of anthropogenic changes in groundwater chemistry have been observed in the aquifer. On the basis of previous research it was shown that the majority of the groundwater components (also indicators of groundwater contamination) do not exceed permissible limits for drinking water. However, in some regions significantly increased concentrations (exceeding the baseline hydrogeochemical setting) of some contamination indicators were observed (Górski 1989, Dragon 2003). These groundwater components have been used for the contamination index calculation.

The formula defining the contamination index (C_d) is (Backman et al. 1998):

$$C_d = \sum_{i=1}^{n} C_{fi}$$

where:

$$C_{fi} = \frac{C_{Ai}}{C_{Ni}} - 1$$

C_{fi} – contamination factor for the i-th component
C_{Ai} – analytical value of the i-th component
C_{Ni} – upper range of natural hydrogeochemical background (Table 1).

Table 1.   The characteristics of contamination indicators for the Wielkopolska Buried Valley aquifer.

Indicator	Min	Max	Average	Median	Standard deviation	Upper range of baseline hydrogeochemical setting	Percent of analyses exceed baseline concentrations
Total hardness (TH) [mval/l]	2.2	10.7	5.8	5.8	1.7	7.0	26
Total Dissolved Solids (TDS) [mg/l]	179	672	385	372	110	450	19
Chloride [mg/l]	7	84	23	15	18.6	20.0	34
Sulphate [mg/l]	0	175	42	33	35.8	40.0	37
Sodium [mg/l]	4.4	22.2	11.0	10.4	4.0	15.0	18

frequency of data set n = 61

The assessment of diffuse contamination is based on the analysis of several basic hydrochemical contamination indicators (Górski 2001). The following parameters have been used for identification of groundwater contamination: chloride, sulphate, sodium, total dissolved solid (TDS) and total hardness (TH). It has been shown that all of these parameters present a steady increase in concentration with time (Dragon 2003). Therefore, all of these parameters can be used as indicators of anthropogenic water contamination (Macioszczyk 1991).

In case of TDS and TH (sometimes also sulphate), the groundwater shows a large spatial variability of concentrations, caused by natural hydrogeochemical zoning. This situation creates difficulties in identifying groundwater chemistry changes due to the influence of Man. Zones of high concentrations caused by natural factors have been identified. However, greater concentration variability of these same parameters is caused by the influence of human activity (Górski 1989). This observation is confirmed by the results of multivariate statistical analysis (Dragon 2002). These parameters were, therefore, used in the present study.

The results of physico-chemical analysis of water sampled from 61 exploited wells were used in the study. The results of this work show the current condition (year 2000) of the groundwater chemistry.

4   RESULTS AND DISCUSSION

The map of anthropogenic groundwater contamination in the WBV aquifer obtained on the basis of the contamination index is presented on Figure 4. The areas with $C_d < 0$ indicate groundwater for which there is no evidence of anthropogenic contamination (concentrations of basic contamination indicators below the upper range of natural hydrogeochemical background). However, the areas with $C_d > 0$ indicate noticeably contaminated groundwater

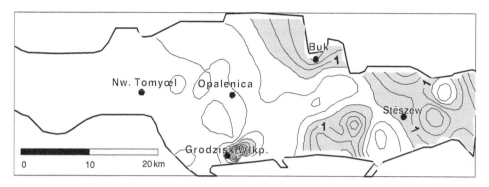

Figure 4.   Groundwater contamination degree estimated on the basic of contamination index (C_d). (The areas with $C_d > 0$ are marked.)

(marked fields on the map), and the areas with $C_d > 3$ indicate the highest degree of groundwater contamination. The chloride concentrations higher than 60 mg/l and sulphate higher than 80 mg/l occur in these areas (these are concentrations significantly above the upper range of natural hydrogeochemical background). The influence of anthropogenic contamination is observed both in urban and rural areas. The most distinct area was found in the eastern part of the study area, where intensive agricultural is practised. It also coincides with the smallest thickness of aquitard, less then 30 m (Figure 3). In the recharge area (the regions of the intertill aquifers – Opalenica town vicinity), where the thickness of the aquitard does not exceed 20 m, the influence of anthropogenic contamination is not observed. This reflects the dominant land use category, which is mainly forests.

The most intensive anthropogenic influences on groundwater chemistry were detected in the region of Grodzisk Wlkp, where the contamination index has $C_d > 7$. In this region the hydraulic contact (buried glacial trough) between the WBV and highly contaminated shallow intertill aquifer was identified. The intensive exploitation of the WBV aquifer (which has a high induced drowndown) and the long-term influence of the unsewered urban area, dating back to the last century, result in intensive contamination of WBV groundwater (Dragon 2004).

The contamination index can be helpful for distinguishing subsets of physico-chemical analysis describing "natural" groundwater chemistry and contaminated groundwater. The sampling points with $C_d < 0$ do not show clear evidence of anthropogenic contamination. However, sampling points with $C_d > 0$ have concentrations of chloride greater than 20 mg/l and sulphide greater than 40 mg/l – thus concentrations greater the upper range of natural hydrogeochemical background, characteristic for confined aquifers in the Great Poland region (Górski 1989).

The difficulties in the interpretation of the anthropogenic changes to the groundwater chemistry may appear when the anthropogenic and natural (geogenic) factors affecting groundwater chemistry overlap. This kind of a situation may occur where concentrations of both TH and TDS as well as sulphate are elevated (Dragon 2003). In addition, *a priori* knowledge of hydrogeologic processes affecting the investigated environment is required for the effective application of contamination index method. It is recommended that this method is used in conjunction with conventional techniques (as in the case of statistical techniques).

It should be emphasized, that the results obtained by the contamination index application are in good agreement with previous studies, performed using multivariate statistical analysis (factor analysis) and conventional techniques (Dragon 2002).

5   CONCLUSIONS

The application of the contamination index method enabled an assessment and a cartographic visualisation of groundwater contamination in a confined aquifer (Wielkopolska Buried Valley aquifer, Poland).

The areas of noticeably contaminated groundwater (concentrations of basic indicators of contamination above the upper range of natural hydrogeochemical background) were estimated and displayed on the map. The contamination occurs mainly in the areas, where the thickness of the aquitard is less than 30 m, and in the most developed regions (intensive agricultural land use, urban unsewered areas).

The data obtained by the contamination index method are consistent with previous studies carried out using multivariate statistical techniques as well as conventional techniques.

The results of an application of the contamination index technique are helpful in distinguishing the vulnerability of the aquifer studied. The noticeably contaminated areas represent the most vulnerable parts of the WBV aquifer.

The contamination index is a very simple method for evaluating and displaying cartographically the degree of groundwater contamination, both in the early stages of anthropogenic contamination (when the majority of ions does not exceed permissible limit for drinking water) and more intense contamination. For an effective application of contamination index method *a priori* knowledge of hydrogeologic processes affecting the investigated environment is required.

ACKNOWLEDGEMENT

The author would like to thank Prof. Dr. J. Górski for his valuable contribution to the topic and important discussions during the study period.

REFERENCES

Backman, B., Bodis, D., Lahermo, P., Rapant, S. & Tarvainen, T. 1998. Application of a contamination index in Finland and Slovakia. *Environmental Geology* 36 (1–2): 55–64.
Dąbrowski, S. 1990. The hydrogeology and conditions of groundwater protection of the Wielkopolska Buried Valley Aquifer (in polish). SGGW-AR Warszawa.
Dragon, K. 2002. Application of the factor analysis for determination of anthropogenic changes in groundwater quality (in Polish). *Polish Geological Review* 50 (2): 127–131.
Dragon, K. 2003. The chemistry of Wielkopolska Buried Valley Aquifer (region between Obra and Warta rivers). PhD Thesis, Adam Mickiewicz University Poznań.
Dragon, K. 2004. Assessment of confined aquifer vulnerability to pollution (Wielkopolska Buried Valley Aquifer, Grodzisk Wielkopolski region, Poland). XVIII National, *Proc. VI International Scientific and Technical Conference "Water Supply and Water Quality"*, Poznań: 267–273.
Górski, J. 1989. The main hydrochemical problems of Cainozoic aquifers located in Central Wielkopolska (Great Poland). *Zeszyty Nauk. AGH:45* Kraków.

Górski, J. 2001. Proposal of anthropogenic contamination evaluation of ground water on the base of chosen hydrochemical indicators. In: *Współczesne problemy hydrogeologii*, T. X., (Bocheńska T. & Staśko S. Ed.) (in Polish). Wrocław: 309–314.

Limisiewicz, P. 1998. The groundwater vulnerability assessment in the chosen basins of the Silesia. *PhD Thesis.* Wroclaw University.

Macioszczyk, A. 1991. Early stages of anthropogenic transformations of groundwater chemical compositions their estimation and interpretation. In: *Współczesne Problemy Hydrogeologii*. T. V. Warszawa – Jachranka (in Polish). SGGW-AR, Warszawa: 254–258.

CHAPTER 8

Application of GIS for presentation of mining impact on change in vulnerability of a Quaternary aquifer

A. Frolik, P. Gruchlik, G. Gzyl, A. Kowalski & K. Kura
Central Mining Institute, Katowice, Poland

ABSTRACT: The hard coal mine "Marcel" in the town of Radlin (Southern Poland) overlaps with the Quaternary Useful Groundwater Aquifer (UGWA) of the Upper Odra River. A layer of quite well isolating loess-like clays and silts covers the aquifer. However, mining activities cause various deformations in the overburden. Particularly, exceeding the certain trigger values of horizontal deformations may result in vertical cracks, the range of which depends on the physical and mechanical properties of the ground. Consequently, the time of vertical migration of contaminants may be significantly shorter, due to preferential flow facilitated by propagating cracks. In particular, if the cracks reach the bottom of the isolating layer, it can lose its protective character at all. The use of GIS for the processing and for the presentation of the data required for the assessment of the mining impact on losing the protective properties of the isolating horizon is presented in this paper.

1   INTRODUCTION

Mining activity is known to cause a lot of adverse effects like damages to buildings, infrastructure, etc. In spite of the threats mentioned above, mining can also pose a risk to groundwater resources. Direct pollution or changes in geochemistry of the bedrock are not the only examples of this kind of threats. Various deformations in the overburden lead to numerous vertical cracks which can appear in the uppermost layers of deep mines overburden. These cracks make it much easier for various pollutants to migrate from the surface into aquifers The use of GIS for the processing and for the presenting of the data required to assess the mining impact on the protective properties of the overburden of aquifers is presented in this paper. This is generally in line with current tendencies reported in scientific papers (Ducci 1999, Fritch et al. 2000, Wu et al. 2004). The chosen example is the Marklowice coalfield, one of numerous coalfields known from the Upper Silesian Coal Basin, the most important coal basin in Poland.

2   BASIC INFORMATION ABOUT THE INVESTIGATED AREA

The Marklowice coalfield (20.5 km^2) is situated in the eastern part of the Marcel coal mine (Figure 1) in Southern Poland.

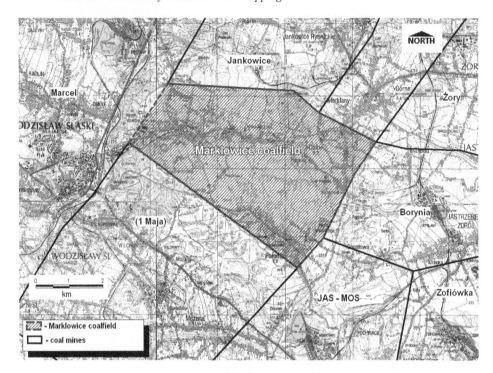

Figure 1.   Marklowice coalfield and surrounding mines.

The flat, plateau-like landscape is cut by deep gorges which reach 20–40 m in depth. The most important streams are Szotkówka and Marklówka and their tributaries.

The geological setting of Marklowice coalfield is the southern part of the Chwałowice syncline between the Michałkowice and the Orłowski reverse faults. Three coal-bearing Carboniferous deposits are known from this area. They are called respectively: Marginal, Saddle and Ruda beds. The upper part of the geological profile consists of Tertiary and Quaternary deposits. The thickness of this overburden increases from West and North-West (50 m) towards East and South-East (370 m). The Tertiary is represented by green and grey Miocene clays of marly and silty character and clayish shales; there are also sandy intercalations of variable thickness (max. 30 m) in the upper part. Quaternary deposits are of an eolian origin (loess-like loams up to 1–20 meters thick) and glacio-fluvial sands, that are of the greatest importance among the whole Quaternary.

There are two aquifers in the investigated area one of Quaternary and one of Carboniferous rocks. These two are separated from each other by impermeable Miocene clays. The Quaternary aquifer is of major importance and is called Useful Groundwater Aquifer (UGWA) of the Upper Odra River and is formed by the sands and gravels of glacio-fluvial origin. This aquifer is almost entirely covered by the layer of loess-like loams. These clays are protecting the Quaternary aquifer from potential pollutants migrating from the surface. The vulnerability of the Upper Odra River UGWA in the investigated area has been described as medium with a vertical migration time (t) of 5 to 25 years and only locally as high (t of 2 to 5 years) or as very low ($t > 100$ years); marginally it is stated as low (t of 25 to 100 years, Różkowski et al. 1997).

The Marklowice coalfield has been mined by three mines: Marcel, Jankowice and 1-Maja (the last one is currently being closed). Marcel coal mine is in operation since 1978 and five coal seams have been mined so far. The thickness of these seams varies from 1.8 m to 4.2 m at the depth of 130–560 m. During 1984 to 2000, 1-Maja coal mine has exploited a dozen of thin (0.8–1.5 m) coal seams underneath the southern part of the Marklowice coalfield at the depth of 280–900 meters. The Jankowice coal mine has exploited the Northern part of the Marklowice coalfield in 24 seams at the depth of 100–450 meters; the seams were 1.6 to 4 meters thick. Longwall mining with breakdown of the ceiling has been the principle method of coal mining and only to some extent the exploited voids were filled hydraulically to support the ceiling.

As a result of the mining activity, the land surface suffered from various deformations of continuous character like subsidence and horizontal displacements as well as their derivations: dip of the ground, curvature and horizontal deformations. There are also discontinuous linear deformations like cracks (closed fractures), fractures, steps, flexures, ground braces and trenches. The exploitation of the Marklowice coalfield resulted also in several subsidence troughs connected among each other. The magnitude of subsidence varies from 2 m in the South, through 7 m in the centre to 20 m in the North. Land surface suffered from ground dips (T) and horizontal deformations (ε) and the area falls into the categories I, II, III, IV and even by larger parts into the category V ($T > 15$ mm/m, $\varepsilon > |\pm 15|$ mm/m) of mining distorted land surfaces (Jędrzejec & Kowalski 1998). There were 56 discontinuous deformations (both simple and complex) detected at the Marklowice coalfield (48 of them at the Marcel coal mine and 8 at the Jankowice coal mine). These discontinuous deformations reveal mainly as fractures and steps.

Both, the Marcel and the Jankowice coal mine, plan to exploit the coal until it is exhausted, which is until 2047 in the Marcel coal mine and until 2035 in the Jankowice coal mine. 205 longwall panels in 21 coal seams are planned for these two mines. Longwall mining is the planned method of exploitation with filling the voids after the exploitation with industrial dust. It is planned to exploit 20 longwall panels of 2–4 meters high each in 9 coal seams at the depth of 280–700 m (Figure 2) from 2003/2004 to 2006.

3 CALCULATIONS

The assessment of the impact of the planned exploitation on the land surface requires the following categories of data:

- field measurements,
- exploitation pattern, and
- the shape and the coordinates of the longwall panels.

It is necessary to digitalize various zones and objects like longwall panels, etc. It is also crucial to formulate the task – the way to calculate deformation indicators as well as the sets of calculation points. After preparation of these data, the forecast of the surface deformations is obtained by the numerical modelling software called Szkody 4.0 (used at the Central Mining Institute, Katowice, Poland (Jędrzejec 2002). The single data sets are gathered and compiled into thematically assembled layers in such a way a kind of database necessary to visualize a certain phenomena is built.

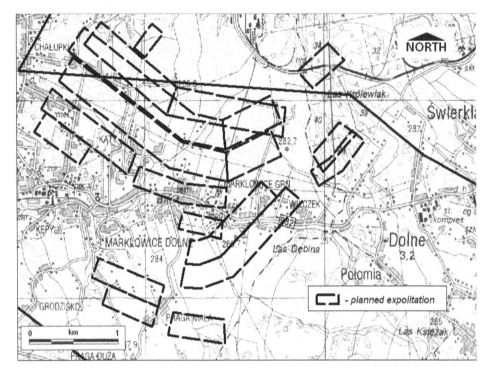

Figure 2.   Planned exploitation of Marklowice coalfield in 2003–2006.

The calculations of maximal horizontal deformations have been performed for the planned exploitation in the Marklowice coalfield from July 2003 to the end of 2006. The general assessment is that the exploitation of deep mines of 100 meters or more below the ground surface results only in continuous deformations. However, this is true only for a low magnitude of surface deformations. Discontinuous deformations (like cracks, etc.) are possible in cases of high-rate deformations in the uppermost ground layer. Therefore, the categories of risk for discontinuous deformations at the surface have been mapped on the base of the abovementioned calculations. The risk categories were described as follows (Figure 3):

- High risk – the forecasted horizontal deformations caused by future exploitation are higher than +9 mm/m,
- Medium risk – the forecasted horizontal deformations caused by future exploitation are from +6 mm/m to +9 mm/m, and
- Low risk – the forecasted horizontal deformations caused by future exploitation are lower than +6 mm/m.

The point at which ground can be no longer deformed in continuous way is called limiting active state. The ground is then deformed along the surfaces of the fractures. The depth of these fractures (h_s) is obtained from the equation [1] (Kwiatek 1982):

$$h_s = \frac{E \cdot \varepsilon}{(1 + v) \cdot v \cdot \gamma} \tag{1}$$

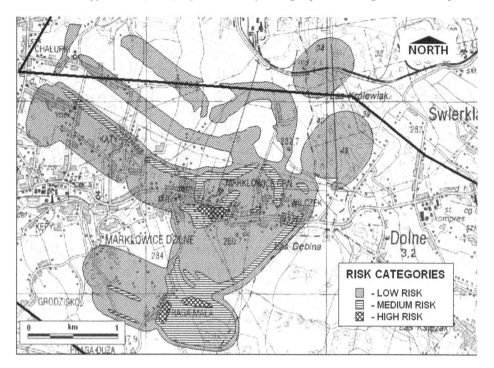

Figure 3.   Risk categories for planned exploitation of Marklowice coalfield.

where: E – ground deformation module,
          ε – horizontal ground deformation,
          ν – Poisson coefficient,
          γ – weight by volume

Laboratory investigations show that the limiting active state can start at horizontal exten-sional deformations with magnitudes of 1.5–3 mm/m for grounds of low cohesion and 6–9 mm/m for grounds of high cohesion. The maximum possible depth of the resulting cracks h_0 is obtained from equation [2] (Kwiatek 1982):

$$h_o = \frac{2 \cdot c}{\gamma} \qquad\qquad [2]$$

where: c – cohesion,
          γ – weight by volume.

As mentioned before, there is a layer of loess-like deposits in the investigated area, which isolates the Quaternary aquifer from the surface. These loess-like deposits are mainly silts, sandy silts, silty loams and loamy sands. Two kinds of geotechnical layers are described among the loess-like grounds with respect to their degree of plasticity I_L: with $I_L = 0.15$ and vivid grounds with $I_L = 0.35$. The later are less abundant, and are mainly found in the val-leys, where the contact with water is more likely. According to Polish Normative PN-B/030-20 (1981), the average values of weight γ and cohesion c are given as 20.4 kN/m^3 and 40 kPa respectively for hard-plastic grounds ($I_L = 0.15$). Furthermore, the average values are

$\gamma = 20.0\,\text{kN/m}^3$ and $c = 25\,\text{kPa}$ for plastic grounds ($I_L = 0.35$). The maximum possible depth of the cracks can be calculated from those data as:

- for hard-plastic grounds:

$$h_0 = \frac{2 \cdot 40}{20.4} = 3.9 \text{ m}$$

- for plastic grounds:

$$h_0 = \frac{2 \cdot 25}{20.0} = 2.5 \text{ m}$$

4   PRESENTATION OF THE RESULTS USING GIS TECHNIQUES

The presence of cracks in the cover formed by isolating loess-like grounds can significantly increase the vulnerability of the Quaternary groundwater aquifer. GIS techniques have been used to present the spatial range of the areas of different levels of risk for increased vulnerability, together with data on the spatial range of Quaternary UGWA and its natural vulnerability. The following thematically assembled vector GIS layers have been created:

- Mining areas – obtained by digitalization of mining areas of Marcel and the neighbouring coal mines;

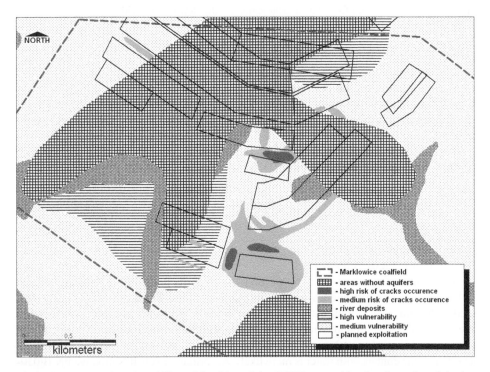

Figure 4.   Increase in vulnerability of the Górna Odra UGWA caused by the planned exploitation of the Marklowice coalfield.

- Stream valleys – obtained by digitalization of the same map (Doktorowicz-Hrebnicki, 1960). Assigned attributes were: 0 – plateau, 1 – valleys;
- Natural vulnerability of Górna Odra UGWA – obtained by digitalization of vulnerability map (Różkowski et al. 1997). Assigned attributes were: 0 – no UGWA, 1 – very low vulnerability, 2 – low vulnerability, 3 – medium vulnerability, 4 – high vulnerability;
- Risk zones – obtained from pre-processing performed by using the Szkody 4.0 software. Assigned attributes were: 0 – no risk of crack occurrence in the ground, 1 – low risk, 2 – medium risk, 3 – high risk.

The attributes assigned to each layer allow to compile the gathered information and to perform the desired query. The queries enable to construct a map of potential of increased vulnerability of Górna Odra UGWA caused by the planned exploitation of the Marklowice coalfield (Figure 4).

The map shows that the majority of zones of high and medium risk of crack occurrence in the UGWA overburden lay in the areas of medium vulnerability (*t* of 5 to 25 years). Since these suspected areas are almost entirely outside of stream valleys, the maximum depth of the cracks (as calculated above) would be about 4 meters. This means, that the overburden of Górna Odra UGWA would lose significantly or even completely its protecting properties. Consequently, an increase in vulnerability of this aquifer is suspected in the areas shown in red and orange at Figure 4. The impact of the planned exploitation is going to be significant, since these areas cover about 50 hectares.

5 CONCLUSIONS

The exploitation pattern of the Marklowice coalfield has been complicated in the past and has resulted in numerous examples of negative impact on the surface. As a result of pre-processing performed by using the Szkody 4.0 software, three categories of risk of crack occurrence in the ground have been established. They are related to the magnitude of deformation which will be caused by the planned exploitation. It was calculated, that the cracks caused by the planned exploitation can be twice as deep in the areas outside the stream valleys as those in the valleys. Since the suspected areas are almost entirely outside of stream valleys, the impact on the vulnerability of this aquifer will be significant. It is estimated, that about 50 hectares of the Górna Odra UGWA overburden are going to loose partially or entirely the protecting character due to the planned exploitation.

REFERENCES

Doktorowicz-Hrebnicki, S. 1960. Mapa geologiczna Górnośląskiego Zagłębia Węglowego bez utworów czwartorzędu. Wydawnictwa Geologiczne. Warszawa.

Ducci, D. 1999. GIS Techniques for Mapping Groundwater Contamination Risk. *Natural Hazards*, vol. 20, no. 2–3, Kluwer Academic Publishers B.V.: 279–294.

Fritch, T.G., McKnight, C.L., Yelderman Jr., J.C., Arnold, J.G. 2000. An Aquifer Vulnerability Assessment of the Paluxy Aquifer, Central Texas, USA, Using GIS and a Modified DRASTIC Approach. *Environmental Management*, vol. 25, no. 3, New York, Springer-Verlag: 337–344.

Kwiatek, J. 1982. Wybrane problemy geotechniki terenów górniczych. Komisja Górnictwa PAN; Katowice.

Jędrzejec, E. & Kowalski, A. 1998. Kategorie terenów górniczych. In: *Ochrona obiektów budowlanych na terenach górniczych* (Kwiatek ed.), Wydawnictwo Głównego Instytutu Górnictwa, Katowice.

Jędrzejec, E. 2002. 32– bitowa aplikacja Szkody 4.0 do prognozowania poeksploatacyjnych deformacji górotworu. *Prace Naukowe GIG*, Seria Konferencje 41, Główny Instytut Górnictwa. Katowice.

POLSKI KOMITET NORMALIZACJI MIAR I JAKOŚCI 1981. Building soils. Foundation bases. Static calculations and design. *Polish Normative No. PN-B/030-20;* Wydawnictwa Normalizacyjne, Warsaw.

Różkowski, A., Rudzińska-Zapaśnik, T. & Siemiński, A. 1997. Map of occurrence, usage, vulnerability and protection of fresh-groundwaters in the Upper Silesian Coal Basin and its margin. Warsaw: Polish Geological Institute.

Wu, Q.,Ye, S., Wu, X. & Chen, P. 2004. Risk assessment of earth fractures by constructing an intrinsic vulnerability map, a specific vulnerability map, and a hazard map, using Yuci City, Shanxi, China as an example. *Environmental Geology*, vol. 46, 1, Springer-Verlag GmbH: 104–112.

CHAPTER 9

A GIS-based DRASTIC vulnerability assessment in the coastal alluvial aquifer of Metline-Ras Jebel-Raf Raf (Northeastern part of Tunisia)

M.H. Hamza,[1] A. Added,[1] R. Rodríguez[2] & S. Abdeljaoued[1]

[1]Université de Tunis El Manar, Faculté des Sciences de Tunis, Département de Géologie, Laboratoire des Ressources Minérales et Environnement, Tunis, Tunisie
[2]Universidad Nacional Autonoma de México, Instituto de Geofisica., Ciudad Universitaria, México D.F., México

ABSTRACT: The alluvial aquifer of Metline-Ras Jebel-Raf Raf is a coastal shallow aquifer located in the Northern east of Tunisia. It occupies a surface of 35 km^2 and is located in the Quaternary parts of the watershed. The rainfall annual average in that region varies between 510 mm to 638 mm.

This paper aims to study the vulnerability to pollution of this aquifer by pesticides and by general pollutants using the DRASTIC pesticides and the DRASTIC standard models. These two models were applied by the GIS technique providing efficient environment for analyses and high capabilities of handling large spatial data. Results are presented as vulnerability maps showing areas with low, moderate and high vulnerability. Lands with high vulnerability to pollution by pesticides occupy 34% of the total aquifer surface and lands with high vulnerability to pollution by general pollutants occupy 10%.

1 INTRODUCTION

Pollution of groundwater supplies represents a serious threat to the population of the coastal watershed of Metline-Ras Jebel-Raf Raf located in the northern east of Tunisia. Protection of the supplies begins with an objective, which is the assessment of the groundwater vulnerability to potential pollution.

Different methods allowing the assessment of the vulnerability to pollution of groundwater are available, but the DRASTIC method (Aller et al. 1987) is the most commonly used in North America, and it becomes more and more used in many other places in the world (Lobo Ferreira & Oliveira 1997; Rodríguez et al. 2001; Ramos & Rodríguez 2003). Aller et al. (1987) agreed that the study of the vulnerability to pollution by pesticides with the DRASTIC model requires the modification of the weights of the parameters comparing to the weights used in the DRASTIC standard model used in the case of general pollutants.

The alluvial plain of Metline-Ras Jebel-Raf Raf is mainly exploited for agriculture, but we must sign that in the last decades, more and more industrial plants have been settled there. This study has been achieved to establish the DRASTIC vulnerability maps to potential pollution by pesticides and by general pollutants in the aquifer of Metline-Ras Jebel-Raf Raf, which could be used to preserve the quality of the groundwater.

In 1999, the vulnerability to potential pollution by general pollutants of the groundwater of Metline-Ras Jebel-Raf Raf has been studied using the DRASTIC standard model (Hamza 1999; Added & Hamza 1999). In this article the DRASTIC parameters has been improved by new data, including net recharge of the aquifer, hydraulic conductivity, geology and soil.

2   METHODOLOGY

The DRASTIC method is used for the evaluation of the vulnerability to potential pollution of the aquifers, by using parametric systems. The common principle of these systems consists in the previous selection of the parameters used in the assessment of the vulnerability. Every parameter is divided into intervals and is affected by a numeric quotation according to its importance in the vulnerability.

There are two types of the DRASTIC model: the DRASTIC standard, applied for general pollutants, and the DRASTIC pesticides. The precision, with which the DRASTIC method permits to distinguish the vulnerable regions, has been verified by many physical and chemical analyses in different climatic regions in the United States, Canada, Mexico, and other countries.

The DRASTIC model is based on the following hypothesis: the potential contamination sources are in the surface of the soil; the potential contamination sources reach the aquifer by the infiltration mechanism; the pollutant has the same mobility comparing to the mobility of the groundwater; and the hydrogeological unit considered has a surface greater than 40 hectares.

The word DRASTIC is an acronym for the parameters used to calculate the index number. The DRASTIC methodology evaluates seven measurable factors (or parameters) for each hydrogeologic setting. Hydrogeological setting is a composite description of all major geologic and hydrogeologic factors, which affect the groundwater movement into, through and out of the area. The factors include: **D**epth to water, net **R**echarge, **A**quifer media, **S**oil media, **T**opography, **I**mpact of the vadose zone, hydraulic **C**onductivity of the aquifer. The most important feature of DRASTIC model is its numerical relative rating and weight system. Each range for each DRASTIC factor has been evaluated with respect to the others to determine the relative significance of each range with respect to their potential pollution. The ratings are from 1 to 10. Each DRASTIC factor has been divided into either ranges or significant media types, which impact on potential pollution. The weighting represents an attempt to define the relative importance of each factor in its ability to affect pollution transport to and within the aquifer. The weight is from 1 to 5.

From these parameters, a DRASTIC index DI or vulnerability rating can be obtained. The higher the value for the DRASTIC index the greater the vulnerability of that location of an aquifer.

$$\mathbf{DI} = \mathbf{D_r}^*\mathbf{D_w} + \mathbf{R_r}^*\mathbf{R_w} + \mathbf{A_r}^*\mathbf{A_w} + \mathbf{S_r}^*\mathbf{S_w} + \mathbf{T_r}^*\mathbf{T_w} + \mathbf{I_r}^*\mathbf{I_w} + \mathbf{C_r}^*\mathbf{C_w}$$

Where **D**, **R**, **A**, **S**, **T**, **I**, and **C** are the seven factors of the DRASTIC method, **w** the weight of the factor, and **r** the rating associated.

Table 1. Weights of the factors in the DRASTIC pesticides and DRASTIC standard models

Factor	DRASTIC pesticides	DRASTIC standard
D: **D**epth to Water	5	5
R: Net **R**echarge	4	4
A: **A**quifer Media	3	3
S: **S**oil Media	5	2
T: **T**opography	3	1
I: **I**mpact of the vadose zone	4	5
C: Hydraulic **C**onductivity of the aquifer	2	3

Table 2. Criteria of the evaluation of the degrees of vulnerability

Degree of vulnerability	DRASTIC index
Low	1–100
Medium	10–140
High	141–200
Very high	>200

The weighting difference between the pesticides and the standard DRASTIC versions (Table 1) is related to the difference in origins of the considered pollutants: in the first version pollutants are inorganic and in the second they are organic. The weight attributed to the parameter soil media is higher in the case of the Pesticide version, 5 as compared to 2 in the standard version. This difference is explained by the fact that the nature of the soil plays a more important role in the adsorption, and neutralisation of pesticides. The same reasoning is applied for the topography parameter where a weight equal to 3 is attributed in the case of pesticides version. In fact, if a land is flat, the pesticides will be leached easily. However, the effect of topography is not too significant on the destiny of inorganic pollutants that is why the weight attributed to this parameter is equal to 1 in the standard version. The other factors were adjusted in the same manner basing on the characteristics and chemical specificities of the pesticides (Aller et al. 1987). In deed, the weight attributed to the parameter vadose zone and the weight attributed to the parameter hydraulic conductivity are quite less elevated in the case of pesticide version as compared to the standard one (4 instead of 5, and 2 instead of 3, respectively).

The values of the DRASTIC index obtained represent the measurement of the vulnerability of the aquifer. These values vary from 26 to 256 in the case of the DRASTIC pesticides version, and from 23 to 226 in the case of the DRASTIC standard version and are classed into four classes corresponding to the different degrees of the vulnerability (Table 2).

3 STUDY AREA

The plain of Metline-Ras Jebel-Raf Raf (Figure 1), is a coastal plain located in the northern east of Tunisia, between the latitudes North 514.411 and 526.215 and the longitudes East 431.657 and 440.351 (UTM Ref). The watershed occupies a surface of $50 \, km^2$ and mountainous border heights exceed rarely 300 m. The alluvial groundwater extends for $35 \, km^2$ and is located between the quotes 0 and 50 m. The rainfall annual average varies

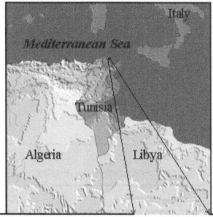

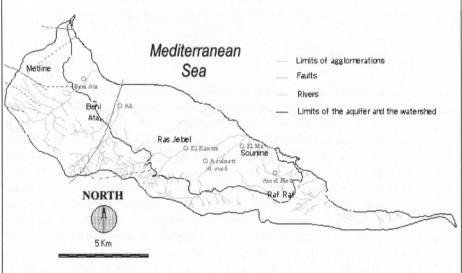

Figure 1.   Study area map.

between 510 mm to 638 mm. The temperature annual average is about 18°C. The main cities and villages in the watershed are: Metline, Ras Jebel, Raf Raf, Beni Ata, and Sounine. The alluvial groundwater plays an important economic role for the agricultural and industrial activities. More than 200 wells are currently exploited in the region and a volume of 1 million m^3 for irrigation.

4   THE DATA

The identification of the hydrogeological units and subunits as well as the assessment of the seven factors of the DRASTIC model requires a good knowledge of the geology, the hydrogeology, the soil media, the topography, and the meteorology in the study area. In the region, many scientific studies were done permitting to determine the DRASTIC factors for the whole territory: climatic study (INM, 1992–2002), geological studies (Burollet 1951;

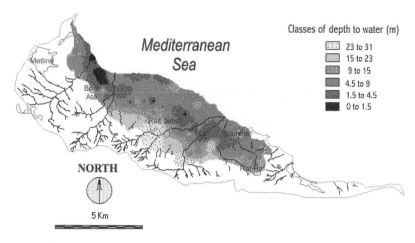

Figure 2.   Depth to water map.

El Ghali & Ben Ayed 2000), geophysical study (Essayeh 1996), hydrogeological studies (Ennabli 1969; DGRE 1992–2002), soil studies (Mansour 1988; Fournet & Mouril 1990; ASB, 1997–2000) and topographical study (OTC 1981).

5   APPLICATION OF THE DRASTIC PESTICIDES AND STANDARD VERSIONS

The GIS system is considered as the more adequate tool used in the application of the DRASTIC model. Two GIS software, ARC/Info and Idrisi, were used in order to apply the DRASTIC standard model in the area. The study area was analysed under a raster (grid) scheme and divided into sub-areas and cells, and each one of them was provided of its appropriate rating value (from 1 to 10). The raster scheme allows the elaboration of thematic maps for each critical factor through the visualisation of zones with different rating values. Then, an aquifer pollution potential index (weight), which is varying from 1 to 5, is assigned to each map. Finally the summation of the seven maps obtained gives us the final DRASTIC values map, that can be classed into four vulnerability classes.

The method used in the elaboration of the database is the following:

– The **depth to water** was obtained from the data of depth to water recorded in 2002, in about 60 wells scattered in the study area (Figure 2)
– The **net recharge** of the aquifer (R) was calculated using the data of the annual precipitation including irrigation (P + Ir) and the data of the annual real evapotranspiration (ETr) (**R = (P + Ir) – Er**) (Fetter 2001). The real evapotranspiration is valued by using the Turc equation (1954) that depends of the annual precipitation P, and of the mean annual temperature T:

$$\text{ETr} = P/[0{,}9 + (P^2/L^2)]^{1/2}, \quad \text{where } L = 3{,}000 + 25T + 0{,}05\, T^3$$

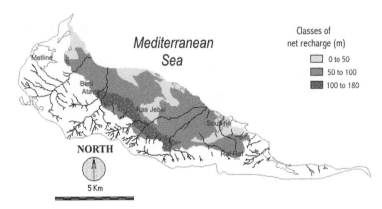

Figure 3.   Net recharge map.

It is important to sign that the amount of water brought by the irrigation from the wells and from the irrigated perimeter of Ras Jebel has been added to the amount of precipitation to calculate the net recharge of the aquifer (Figure 3).

This method used in 2003 to calculate the net recharge of the aquifer was different from that used in 1999. In fact, in the previous study the net recharge was calculated using the soil water balance method (Thornthwaite 1948). This includes the following parameters: the monthly precipitation, the artificial recharge, the superficial run-off, the ground moisture content, the evaporation and the plant transpiration. The following algorithm is used in that method:

$$R = \sum_{i=1}^{12}(P_i + Ir_i - R/O_i - \Delta ST_i - ETR_i)$$

where: P_i is the rainfall in the ith month [mm/month]; Ir_i = artificial recharge in the ith month [mm/month]; R/O_i = superficial run-off in the ith month [mm/month]; ΔST_i = variation of the ground moisture content in the ith month [mm/month]; and ETR_i = actual evapotranspiration (i.e. evaporation and plant transpiration) in the ith month [mm/month].

This equation is not taken into consideration in the following article because of lack of information for the whole area. The Turc's method used in that study presents although some limitations, but it was considered appropriated to the study.

– The data of the **aquifer media** (Figure 4) is determined in two ways. In the first way, we used the data related to the boring logs done by Ennabli in 1969, then we evaluated and updated the thickness of the aquifer zone according to the depth to water variation noted between 1969 and 2002. In the second way, we used the aquifer hydraulic conductivity data (see below), particularly in the areas where boring logs do not exist, to estimate the aquifer lithology.
– The **soil** data is based on soil map (scale 1/12,500) (Mansour 1988; Fournet & Mouri 1990), and on some other local studies (ASB, 1997–2000) (Figure 5). In the previous study of 1999, only a map of soil use was available with a scale of 1/200,000.
– The **surface acclivity** (slope) was calculated by using the three topographic maps covering the study area (scale of 1/25,000) (Figure 6).

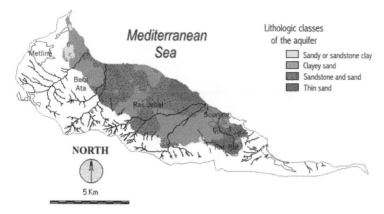

Figure 4.   Aquifer media map.

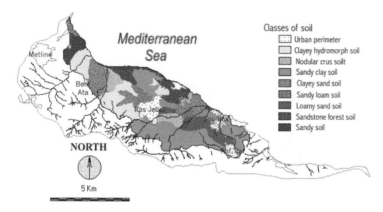

Figure 5.   Soil media map.

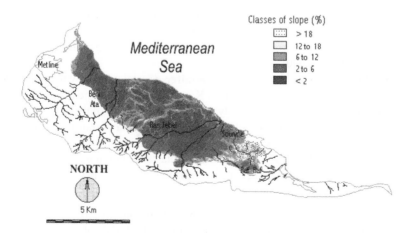

Figure 6.   Surface acclivity (slope) map.

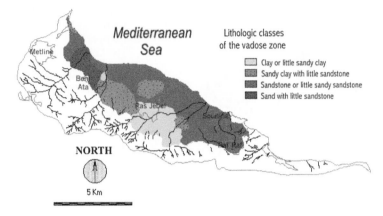

Figure 7.   Impact of the vadose zone map.

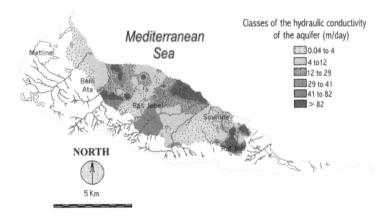

Figure 8.   Hydraulic conductivity map.

– The data related to the **lithology of the vadose zone** were determined by applying a correlation between the different existing bore logs and also by using the geology data of the old geology map (Burollet 1951), which was not covering all the study area, and the new one (El Ghali 2000) (Figure 7).

– Finally the **hydraulic conductivity of the aquifer** was calculated by using the data of transmissivity and thickness of the aquifer: $K = T/b$; where K is the hydraulic conductivity of the aquifer, T is the transmissivity and b is the thickness of the aquifer (Figure 8). As opposed to the study achieved in 1999, on which the data of transmissivity was only available for 20 points covering less than the half of the groundwater surface, we were able in the present study to use a transmissivity data related to more than 70 points covering the totality of the groundwater surface, and that allowed a better determination of the hydraulic conductivity of the aquifer.

To obtain the final vulnerability maps to potential pollution by pesticides (Figure 9) and by general pollutants (Figure 10), we multiplied, using the Idrisi software, every map of each critical factor by the value of its relative weight, and we summed the seven numerical

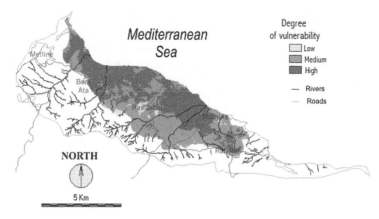

Figure 9.   DRASTIC pesticides map (2003).

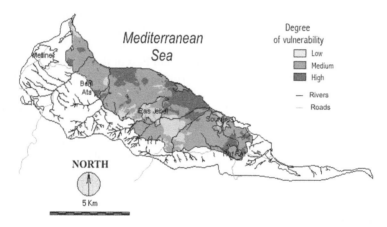

Figure 10.   DRASTIC standard map (2003).

maps obtained. The final maps obtained, which represents the maps of indexes of vulnerability (DI), classified the aquifer into three different degrees of vulnerability.

6   RESULTS AND DISCUSSION

The final DRASTIC pesticides and standard maps obtained show three classes of vulnerability in the studied area: low, medium and high. These maps were achieved at a better scale (1/25,000) than the map prepared in 1999 (scale: 1/100,000). Both maps show three classes of vulnerability: low, medium and high. They are considered as operative maps in the classification of Civita (1994). Civita classed the vulnerability maps into 4 classes according to the denominator of their scales (DS): Orientative maps for DS ≥ 500,000; Schematic maps for 100,000 ≤ DS < 500,000; Operative maps for 25,000 ≤ DS < 100,000 and Special maps for DS < 25,000.

The obtained results related to the vulnerability to pesticides show the high extension of the zones with medium vulnerability (64%) comparing to the others zones of the groundwater.

The zones with high vulnerability occupy 28%, and the zones with low vulnerability occupy 8% of the total surface of the groundwater. The zones with high vulnerability are located mainly in the coastal areas between Essafia in the West and the village of Sounine in the East. We can also find them in the town of Raf Raf and in the South of the village of Ras Jebel. The high vulnerability in these zones is mainly related to the low values of depth to water, the high permeability of the soils and the high permeability of the vadose zone materials, which are mainly constituted by sandstone or by intercalations of sand and sandstone. The moderate vulnerability areas are located in a vast zone located in the East, West and South of the town of Ras Jebel. Other occurrence is related to the area of the village of Sounine, at the crossroads of the Ras Jebel-Metline and Metline-El Alia highways, and in the zone of Ras Zebib. Finally, the areas with low vulnerability are located in the extreme East and West of the town of Ras Jebel, in the region of El Koudiet, in the West of the town of Ras Jebel, and in some restricted zones in the village of Sounine and in the region of Ras Zebib.

From an other part, the obtained results related to the vulnerability to general pollutants show equally the high extension of the zones with medium vulnerability (64%) comparing to the others zones of the groundwater. The zones with high vulnerability occupy 19% and are mainly located in two extended coastal regions located in the East and in the West of the town of Ras Jebel, in some other little areas located in The West of the town of Metline, in the North of the town of Raf Raf, and in the South East of the village of Beni Ata. The zones with low vulnerability which occupy 17% of the total surface of the groundwater are mainly located in the region of Ras Zebib, in the South of the town of Ras Jebel, and in its South East and South West borders, and in the village of Sounine and its Southern borders.

Finally we have to notice that in general, in our study area, the areas located in the beds of the rivers crossing high vulnerable zones are characterized by a permeable vadose zone, and that can increase the risk of the pollution by pesticides and by general pollutants in these areas.

7   CONCLUSION

The application of the DRASTIC pesticides model shows that more than the half of the aquifer area (64%) presents a medium vulnerability to pollution by pesticides. The zones with high vulnerability cover also an important surface (28%). These results show that the groundwater of the plain of Metline-Ras Jebel-Raf Raf is threatened by the pollution by pesticides which are usually used in the agricultural field in the region. From another part, the greatest part of the aquifer presents a medium vulnerability to pollution by general pollutants (64% of the study area). The zones with high vulnerability to pollution occupy only 19%. Nevertheless, this study deserves to be confirmed by doing analyses of pesticides and general pollutants in the groundwater.

REFERENCES

Added, A. & Hamza, M.H. 1999. Evaluation of the vulnerability in Metline aquifer (Northeast of Tunisia). ESRI user conference. San Diego, USA.
Aller, L., Bennett, T., Lehr, J.H., Petty, R.J. & Hackett, G. 1987. DRASTIC: a standardized system for evaluating ground water pollution potential using hydrogeological settings. US-EPA Reports, 600/2-87-035, Washington DC, USA.

ASB: Arrondissement des Sols de Bizerte. 1997–2000. Comptes rendus d'études pédologiques établis à Metline, Ras Jebel et Raf Raf.

Burollet, P.F. 1951. Etude géologique des bassins Mio-Pliocènes du Nord Est de la Tunisie. *Ann. Mines et Géol.* n°7, 82 p + annexes + cartes 1/50.000.

Civita, M. 1994. La carte della vulnerabilità degli acquiferi all'inquiamento: Teoria e pratica. Pitagora editrice, Bologna: p. 325.

DGRE: Direction Générale des Ressources en Eau. 1992–2002. Annuaires piézométriques de la Tunisie.

El Ghali, A. & Ben Ayed, N. 2000. Carte géologique au 1/50,000 de Metline. Publications du Service Géologique de Tunisie.

Ennabli, M. 1969. Etude hydrogéologique de la plaine de Ras Jebel. Rapport interne DGRE, réf 7/57: p. 134 + annexes.

Essayeh, F. 1996. Apport de la méthode de prospection électrique à l'étude des problèmes d'intrusion marine dans la plaine de Metline Ras Jebel Raf Raf. DEA, Univ Tunis II, FST: p. 100 + annexes.

Fetter, C.W. 2001. Applied Hydrogeology, 4th ed., Charles Merrill Pub. Co. Columbus Ohio: p. 598.

Fournet, A. & Mouri, A. 1990. Reconnaissance pédologique des périmètres d'irrigation de Beni Ata et de Chaab Eddoud. Rapport interne, Direction des Sols.

Hamza, M.H. 1999. Etude de la vulnérabilité à la pollution potentielle de la nappe de Ras Jebel par les systèmes d'information géographique. DEA, Univ. Tunis II: p. 102.

INM: Institut de la Météorologie Nationale, 1993–2003. Tableaux climatologiques mensuels, station Bizerte-Sidi Ahmed.

Lobo Ferreira, J.P. & Oliveira, M.M. 1997. DRASTIC Groundwater Vulnerability Mapping of Portugal. In Groundwater: An Endangered Resource. *Proceedings of Theme C of the 27th Congress of the International Association for Hydraulic Research. S. Francisco, USA.*

Mansour, H. 1988. Carte pédologique de Beni Ata, Ras Jebel et Raf Raf au 1/12.500, Publ. Direction des sols de Tunisie.

OTC: Office de la Topographie et de la Cartographie. 1981. Carte topographique de la Tunisie au 1/25.000. Feuilles de Metline S.O, Metline S.E et Ghar El Melh N.E.

Ramos, J.A. & Rodríguez, R. 2003. Aquifer Vulnerability mapping in the Turbio River valley, Mexico: A validation study. *Geofisica International.* Vol. 42, 1, 141–156.

Rodríguez, R., Reyes, R., Rosales, J., Berlin, J., Mejia, J.A. & Ramos, A. 2001. Estructuracion de mapas tematicos de indices de vulnerabilidad acuifera de la mancha urbana de Salamanca Gto., Municipio de Salamanca. CEAG, IGF-UNAM. Technical Report, p. 120.

Thornthwaite, C.W. 1948. An approach toward a rational classification of climate. *Geographical Review*, New York, vol.38, 1: 55–94.

Turc, L. 1954. Le bilan d'eau des sols : Relations entre les précipitations, l'évaporation et l'écoulement. *Ann. Agron.*, 5: 491–595.

CHAPTER 10

The changes of groundwater quality of the "Czarny Dwór" intake as a result of the aquifer vulnerability

B. Jaworska-Szulc, B. Kozerski, M. Pruszkowska & M. Przewłócka
Technical University of Gdańsk, Faculty of Hydro and Environmental Engineering,
Department of Hydrogeology and Engineering Geology, Gdańsk, Poland

ABSTRACT: The "Czarny Dwór" groundwater intake is one of the main sources of water for Gdańsk. It is situated on the Marine Terrace and exploits mainly Quaternary aquifer, but also Tertiary and Cretaceous. The water is generally of good quality although some undesirable changes were observed in the Quaternary aquifer during almost 40 years of exploitation. The changes are caused by developing urbanization, former high pumping rates and natural conditions such as vicinity of the sea and vulnerability of the aquifer. They are expressed in an increased amount of Cl, SO_4, N/NH_4 ions and total hardness.

1   INTRODUCTION

The "Czarny Dwór" groundwater intake is one of the most significant well fields to the Gdańsk water supply from almost 40 years. The line of wells, 3,200 m long is situated 600–1,500 m from the sea shore. The abundance of water in this area was noticed already at the beginning of XXth century by German hydrogeologist Thiem. The development of the surrounding urban area though, together with the location close to the shore line, has brought some threats to the water extracted from the Quaternary aquifer. Groundwater risk assessment and protection of the water was the aim of different investigations and studies carried out for many years.

From 2002, a team from Royal Technical University of Stockholm together with Polish hydrogeologists and students from Technical University of Gdańsk has also contributed in those studies. The first year of co-operation was concentrated on the assessment of salinization from the Baltic Sea, mainly using geophysical methods and also on the influence of the allotment gardens situated in the area of the intake.

2   HYDROGEOLOGICAL CONDITIONS

There are three aquifers containing water of good quality on the Marine Terrace: Quaternary, Tertiary and Cretaceous.

The Cretaceous aquifer is connected with widespread hydrogeological structure called Gdańsk Artesian Basin. It consists of fine – grained, quartz – glauconite sands of

Coniacian and Santonian. The deposits, 150 m thick occur on the depth of about 150 m and their transmissivity is 30–350 m²/h. The recharge zone of the aquifer includes the area of the Kashubian Lake District where the piezometric surface is approximately 150 m a.s.l. The discharge zone of the basin is located on the terrains of the Vistula River Delta and the Bay of Gdańsk together with the Marine Terrace. The head drops to 15 m a.s.l in natural conditions.

The Tertiary aquifer in this area consists of sands, in some places with gravel and phosphate concretions of Oligocene. The roof of water bearing strata is lies about 60–75 m a.s.l and the transmissivity is 12 – 36 m²/h. The recharge zone is on the Kashubian Lake District where the original piezometric surface occurs at 120 m a.s.l. and the zone of discharge is the Bay of Gdańsk, the Marine Terrace and the Vistula River Delta. The natural head of the piezometric level is about 8 m a.s.l.

The Quaternary aquifer is composed of fluvioglacial sands and gravel originating from Middle Polish glaciation. They are covered by Holocene sands of outwash cones and nearer to the sea by marine sands. Those deposits are interbeddeed with silts and silty sands in the part of area adjoining the sea (Figure 1). The transmissivity of the aquifer is 5–200 m²/h. In natural conditions the water table, or locally the piezometric surface, varied from 2 to 4 m a.s.l. In some parts of the area the Quaternary aquifer is covered by peat but in a considerable part of the region there is no confining layers, so it is rather exposed to contamination. The recharge is mainly lateral, by a direct inflow from the Kashubian Lake District, and also in some extent by atmospheric precipitation and the ascension of water from Tertiary aquifer. In natural conditions the groundwater table close to the sea was about 2 m a.s.l., which means that the discharge was taking place a few kilometres from the seashore in the Baltic Sea (Kozerski 1988).

The natural circulation of groundwater was disturbed by exploitation. The most intensive abstraction took place in 1984–1985 when in the centre of the cone of depression the water table was lowered to approximately 3.5 m b.s.l. At present, the water table is rising

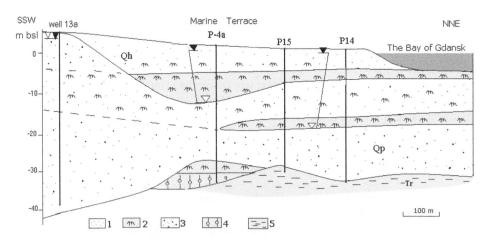

Figure 1.   The hydrogeological crossection through the Czarny Dwór groundwater intake. 1 – sand, 2, 5 – silt, 3 – sand with gravel, 4 – boulder clay.

in all three exploited aquifers. The groundwater level in Quaternary aquifer locally reaches ground level.

3 THREATS TO THE GROUNDWATER QUALITY AND THE METODICS OF INVESTIGATIONS

One of the threats to the exploited Quaternary water is close proximity of the sea. The intensive exploitation in the period of 1965–1985 led to an extensive cone of depression spreading under the sea bed. The intrusion of salt water to the Quaternary aquifer was provoked in the 80's and 90's. Another factor unfavourable for the quality of groundwater is the lack of confining layers in the upper part of the aquifer in a 1 km wide area surrounding the line of wells, which means exposure to potential contamination from the surface.

An obvious threat to the well field is also a development of the surrounding urban area. The risk is connected with expending heavy traffic road net, development of sewage system and other underground installation, existence of such objects as waste disposal plant, hospital, transport base, petrol stations, garages and factories and also allotment gardens which had existed for 20 years in the closest surroundings of the intake. The allotments may constitute a risk of pollution due to excessive use of fertilizers and insecticides. It was also stated in the past that on the area of the gardens there existed some incorrectly constructed shallow wells used as cesspools or compost tanks. Because of the risks a decision to remove the allotments was taken recently. From last year tidying up works are carried on and the action to improve the recreational value of this part of the area is taken, together with land reclamation works necessary because of the present high groundwater level.

In April 2002, when the field work was undertaken, there was disturbed ground from the destruction of the allotments, arbours, fences and with the rising groundwater causing surface ponding. In the days 24th–27th of April some samples were taken in order to make chemical analysis in the laboratory. These were 9 samples of groundwater taken from piezometers located between the line of wells and the sea shore, 7 samples of surface water from drainage ditches and floods and 5 samples of soil water taken from lysimeters. In the samples taken to laboratory the following parameters and ions were measured: colour, alkalinity, Cl^-, SO_4^{2-}, N/NO_3^-, Ca^{2+}, Mg^{2+}, Na^+, K^+, Fe^{2+}, Mn^{2+}. Electrical conductivity and pH was measured in the field. The points of sampling and the location of the intake is shown on Figure 2. The Figures 3 and 4 shows concentrations of chosen parameters in all three types of water.

In order to detect possible salt water intrusion in the study area, a geophysical method – Vertical Electrical Sounding (VES) – was applied. Thanks to differences in resistance between salt and fresh water it is possible to detect boundaries of salt water occurance using geophysical methods. The profiles were made at eight different locations (Figure 2) and the Schlumberg's electrodes system was adopted in the measurements.

4 THE RESULTS OF GEOPHYSICAL MEASUREMENTS

Table 1 shows that in all the profiles (except for the measurements obtained at the shore line) there is a clear reduction of the apparent resistivity values at the depths of 1.5–6.8 meters, which is due to the level of groundwater table. At the depths 6.8–10 or

10–15 meters the profiles show in some cases an increase of resistivity values, which may be explained by the fact that there is a change in the layers from sand to sand mixed with silt at these depths. It is visible especially in the zone around points P-14–P-15. The values measured at the shore line are very low because they show the resistivity of sea water.

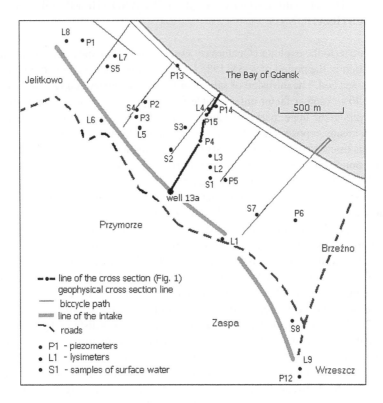

Figure 2.   The scheme of the intake with localization of sampled points.

Table 1.   Values of the apparent resistivity at the 8 measured points.

Depth [m]	well 13a	P4a	P15	P14–P15	P14 path	Bicycle	Beach	Shore line
				resistivity [Ωm]				
1.5	3,638	173	880	663	1,652	3,146	588	13
2.2	3,758	119	649	621	886	2,530	654	12
3.2	3,580	86	455	422	920	1,671	594	11
4.6	2,661	78	256	239	296	920	453	11
6.8	1,445	81	136	98	134	325	260	13
10	549	78	103	94	56	157	157	16
15	353	78	177	113	85	177	127	21
22	91	91	106	85	91	91	182	23
32	45	97	109	106	97	97	193	32
46	106	126	146	233	798	333	1,197	40
68	160	203	65	–	–	857	1,743	–
100	628	7,854	361	–	–	198	2,199	–

The main result is that the measurements do not show any salt-water intrusion into the aquifer.

5   THE RESULTS OF CHEMICAL WATER ANALYSIS

The analytical analysis on the samples were performed in the laboratory of Royal Technical University in Stockholm. Figures 3 and 4 show the results concerning SO_4^{2-}, Cl^- and Na^+. All the measured ions and compounds are compiled in Table 2, where the minimum and maximum values are shown in the numerator and the average value in the denominator. All kinds of waters are characterised by high and variable concentration of sulphates but especially surface water ($31–610\,mg/dm^3$). This can be explained by both natural factor and human activity. Sulphates can be formed naturally in an aeration zone in the presence of organic matter. In this case the organic matter is present mainly in peat covering more or less continuously the exploited aquifer. As to anthropogenic origin, one must keep in mind the recent history of the area. The activity in the allotment gardens included probably using different fertilizers such as potassium sulphate or dung and compost. Besides a heating plant using coal existed for many years close to the intake. Sulphur compounds emitted to the atmosphere could have infiltrate to the ground with precipitation where, depending on the red-ox conditions, they were present in different form. Coal burning is one of serious sources of sulphates in shallow groundwater (Witczak & Adamczyk 1995).

The influence of the former allotment gardens is visible mainly in the surface water composition. One of examples can be an increased concentration of potassium (max $40\,mg/dm^3$) and the other – nitrate nitrogen (max $3.59\,mg/dm^3$, average $1.33\,mg/dm^3$). Those ions occur

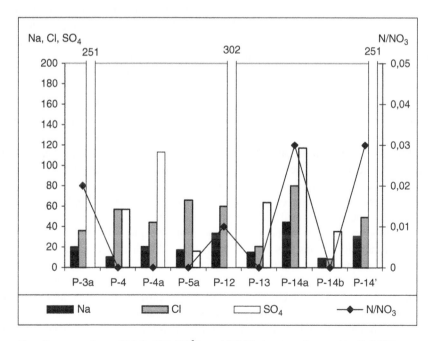

Figure 3.   Concentrations of Na^+, Cl^-, SO_4^{2-}, and N/NO_3 in groundwater (April, 2002).

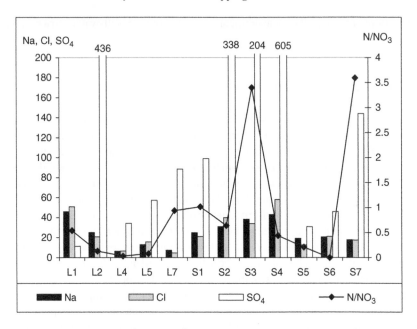

Figure 4.   Concentrations of Na^+, Cl^-, SO_4^{2-}, and N/NO_3 in surface and soil water (2002).

in typical amounts in the groundwater though. This is quite understandable taking into consideration the current hydrodynamic conditions in the area in question. The hydraulic head in the abstracted aquifer is higher than in the subsurface water. Consequently the contaminants from the shallow water don't percolate into the main aquifer. The situation could changed if abstraction increases yield of the intake. This situation took place in the past.

Figures presented in the Table 2 also shows wide variations of Fe^{2+} and Mn^{2+} amounts in soil and surface water. Although the concentrations are sometimes very high (max 75.4 mg/dm^3 in soil water and max 36 mg/dm^3 in surface water) it should not be regarded as an anomalous phenomenon. Such high amounts of those ions are sometimes reported in the presence of organic matter in sediments (Ratajczak, Witczak, 1983).

Figures 5 and 6 present variations of SO_4^{2-}, Cl^-, N/NO_3 and N/NH_4 in the wells of the intake in different periods of time, as a background for the investigations conducted in April 2002. The chosen ions can be admitted as indicators of human influence on the chemistry of groundwater. The sequence of numbers on the axe of abscissae responds to their actual location in the field. The periods of time are: 1978 (more or less natural or slightly changed composition of groundwater), 1992 (the year of highest amount of Cl^- ion in piezometers as a result of overexploitation) and 2002 (current situation).

The graphs show changes in chemical composition of exploited water. The lowest concentration of majority of ions was in 1978. It is also visible that the amount of some compounds, especially ammonia nitrogen and sulphates varies in wide limits. Most significant changes concern the S-E part of the intake where such objects as sewage disposal plant, hospital, big buses parking place, highway, allotment gardens are concentrated in close vicinity of the wells. Those variations goes together with an increased amount of Cl^- ion, which rather shouldn't be admitted in this part of area as the effect of salt water intrusion, but as the effect of human activity.

Table 2. Minimum, maximum and average values of selected parameters in waters taken from piezometers, lysimeters and surface water.

Parameter	Piezometers	Lysimeters	Surface water	Parameter	Piezometers	Lysimeters	Surface water
conduct. [µS/cm]	409 – 1,085 / 664	268 – 994 / 574	431 – 1,307 / 835	HCO_3^- [mg/dm^3]	145 – 285 / 215	138 – 233 / 190	120 – 305 / 226
pH	6.87 – 7.59 / 7.27	6.64 – 7.33 / 6.93	6.7 – 8.05 / 7.15	Cl^- [mg/dm^3]	8.5 – 80 / 46.8	4.7 – 50.9 / 19.7	12.8 – 58 / 29.2
Na^+ [mg/dm^3]	9.1 – 44.5 / 22.3	6.43 – 46.0 / 19.6	17.8 – 42.9 / 27.9	N/NO_3 [mg/dm^3]	0 – 0.03 / 0.01	0.029 – 0.94 / 0.344	0 – 3.59 / 1.33
K^+ [mg/dm^3]	2.3 – 5 / 4.04	0.57 – 5.7 / 2.55	0.65 – 43 / 11.1	SO_4^{2-} [mg/dm^3]	15.9 – 302 / 134	11.3 – 436 / 125.5	31 – 605 / 210
Ca^{2+} [mg/dm^3]	58 – 155 / 97.4	40 – 160 / 103	63 – 227 / 130	Fe^{2+} [mg/dm^3]	0.91 – 6.8 / 2.45	0.04 – 75.4 / 28.19	0.22 – 36 / 11.03
Mg^{2+} [mg/dm^3]	9.8 – 22.8 / 16.7	1.5 – 19.1 / 10.1	4.6 – 14.8 / 10.1	Mn^{2+} [mg/dm^3]	0.07 – 0.327 / 0.167	0.004 – 0.87 / 0.389	0.186 – 1.16 / 0.576

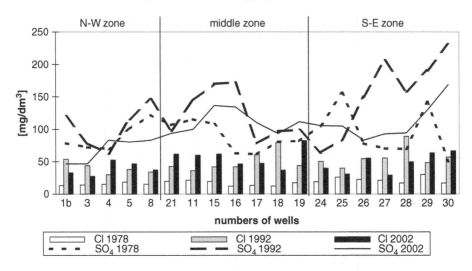

Figure 5.   The changes of Cl⁻ and SO_4^{2-} concentration along the line of wells in different periods of time.

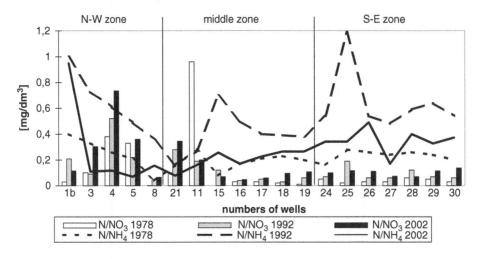

Figure 6.   The changes of N/NO_3 and N/NH_4 along the line of wells in different periods of time.

6   CONCLUSIONS

The results of geophysical investigations show that there is no longer any salt water intrusion into the Quaternary aquifer. In all the profiles in the coastal area only fresh waters were found.

The chemical analyses indicate the human influence on the groundwater composition, especially in the S-E part of the abstracted area where some potential sources of contamination are in close proximity. Those changes do not prevent the abstracted water from being used for drinking purposes. They only show some processes that are taking place in the aquifer. A distinctive feature of all kinds of waters is the wide variation of ammonia nitrogen, nitrate nitrogen and sulphates, which can be related to the method of the land use. The influence of the allotment gardens is clearly visible in the chemical composition of surface and soil water (increased amount of sulphates, potassium and nitrate nitrogen) but not in groundwater. This would be expected in the current hydrodynamic conditions, when the exploited aquifer shows a higher hydraulic head than the subsurface water. The situation can be changed in case of increased exploitation.

REFERENCES

Kozerski, B. 1988. Warunki występowania i eksploatacja wód podziemnych w gdańskim systemie wodonośnym. Mat. IV Sympozjum *Aktualne Problemy Hydrogeologii*, Gdańsk, Wyd. Instyt. Morskiego, cz.1: 1–20.

Ratajczak, T. & Witczak, S. 1983. Mineralogia i hydrogeochemia żelaza w kolmatacji filtrów studziennych ujmujacych wody czwartorzedowe. *Zesz. Nauk. AGH*, z.29. Kraków.

Witczak, S. & Adamczyk A. 1995. Katalog wybranych fizycznych i chemicznych wskaźników zanieczyszczeń wód podziemnych i metod ich oznaczania. PIOS, Warszawa: p. 584.

CHAPTER 11

Groundwater vulnerability to contamination in the central part of Vistula River valley, Kampinoski National Park, Poland

E. Krogulec
Warsaw University, Faculty of Geology, Institute of Hydrogeology and Engineering Geology, Warsaw, Poland

ABSTRACT: Evaluation of groundwater vulnerability to contamination was conducted in the 650 km^2 middle part of Vistula River valley, Kampinoski National Park (KNP) area, Poland. Two evaluation methods were used: the U.S. EPA DRASTIC method and the time migration method of conservative contamination from the terrain surface (percolation time through unsaturated zone). The resulting maps, obtained from U.S. EPA DRASTIC, of groundwater vulnerability show that area of study is generally characterized by medium and medium high natural vulnerability. The time migration is generally included in the range of 30 days to 3 years.

1 INTRODUCTION

Groundwater vulnerability should be considered during planning and designing development of an area, especially if it is located where it can pose a threat to groundwater. Groundwater vulnerability is a concern even in recreation areas such as national parks or landscape parks. Such areas, including Kampinoski National Park in Poland, are not isolated from "outside effects", and may be a place of constant by the neighbouring inhabitants. Even thou exists so called protection law of areas like Kampinoski National Park, we have to mention that they are not isolated from "outside effects", and sometimes outright on the contrary they are a place of constant or temporary stay for the neighbouring inhabitants. Because methods for evaluation of groundwater vulnerability to contamination are not standardised, it is difficult to compare results and conclusions between studies potential impacts.

Kampinoski National Park (area – 385,44 km^2) with its buffer zone (area – 385,88 km^2) is a UNESCO Biosphere Reserve, located where four tributaries: the Bug, Narew, Wkra, Bzura Rivers, merge with The Vistula River. According to The Europe Council (Centre Naturopa 1998) the valleys of these rivers are ecological corridors. The Vistula River valley the Kampinos Forest are especially recognized as important ecological areas in Europe.

KNP is situated within the river valley that includes the suburbs of Warsaw, a city of nearly 2 million people (Figure 1). The area is characterized by a diversity of morphology, hydrogeological conditions, geology and variability of vegetation as well as infrastructure development. The biggest municipal waste disposal site in the region is located in the area although it is not in operation. The terrain includes considerable forest and naturally swamp areas.

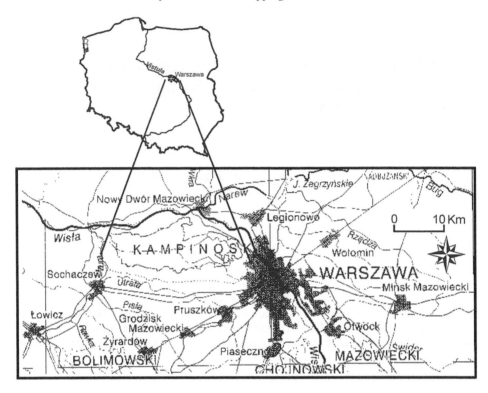

Figure 1.   Location of the Kampinoski National Park (Poland).

2   HYDROGEOLOGICAL CONDITIONS OF THE KNP REGION

Kampinoski National Park and its surroundings are located in the central part of the Vistula River valley. This valley flood plains and a river flood terrace (Kampinos Terrace) of The Vistula and Bzura Rivers. On the Kampinos Terrace, one aquifer occurs with a thickness of 10 to 50 metres, it is composed of varied, fine-grained sands, in some places silty. The groundwater table has an unconfined character. Monitoring of groundwater and surface waters in the KNP area, conducted since 1988, enabled evaluation of the depth of the water table (Krogulec 2001a, 2001b), a fundamental criteria used in evaluation of groundwater vulnerability to contamination. During the period between years 1998 and 2002, the water table was at a depth of 0.30 to 5.52 meters below terrain level.

Aquifer hydraulic conductivity values were determined by statistical analysis of hydrological data obtained from approximately 980 wells within the study area (Krogulec 2003) and from archival data and numerical models (Krogulec 1997a; Krogulec 1997b). Average hydraulic conductivity value for the aquifer sediment was determined from pumping test results to be 47.7 m/d (range of 1.2 m/d to 89.6 m/d).

Groundwater recharge takes place almost exclusively as a result of infiltrating precipitation but a second recharge source comes from a lower aquifer. The lower aquifer only affects the south part of Kampinos Terrace. During high spring runoff, water from the Vistula River may infiltrate the aquifer. Water also infiltrates from the Bzura River in the south-west part study area.

Aquifer drainage on the Kampinos Terrace takes place indirectly with the help of rivers and canals, and partially with evapotranspiration processes in the swampy areas. The Vistula River has the strongest drainage character and is the regional drainage base. Similarly an important role in drainage of aquifer fulfils Bzura, mostly in its lower section. Smaller streams in KNP drain the aquifer to a considerably lesser degree. Drainage of the aquifer also takes place through production of groundwater in many points in the study area.

3   GROUNDWATER VULNERABILITY

In this article groundwater vulnerability to contamination is understood to be a natural characteristic of the aquifer system, it describes the risk of migration of hazardous substances from the surface to the aquifer. Intrinsic vulnerability is determined only by hydrogeological conditions (recharge conditions, discharge, formation conditions including degree of groundwater isolation). Specific vulnerability also takes into consideration type of hazardous substance, its amount, and its location with respect to the aquifer (Duijvenbooden, Waegening 1987; Limisiewicz 1998; Vrba, Zaporozec (eds.) 1994; Witczak, Żurek 1994).

For central area of Vistula River valley (KNP region and its surroundings) natural vulnerability of groundwater was determined with application of two methods:

– U.S. EPA DRASTIC Model
– migration time of conservative contamination from the terrain surface (percolation time through unsaturated zone).

3.1   *DRASTIC model*

One of the most widely used groundwater vulnerability methods is DRASTIC, developed by the United States Environmental Protection Agency (EPA) as a method foe assessing groundwater pollution potential (Aller et al. 1987). In the DRASTIC method specific criteria including: depth to the water level, effective infiltration, aquifer media, type of soil, topography, impact of vadose zone, and the hydraulic conductivity of aquifer, are assigned different degrees of importance on a scale of 1 to 5. Each criterion also possesses suitable order of value of the used coefficient and its credited with a rank, in other words a rank on a 1 to 10 scale (Table 1). Vulnerability index IPZ_Σ that is the sum of the multiplication of variable rank and weight of individual criterions is the final evaluation of groundwater vulnerability.

$$IPZ_\Sigma = \sum_{n=1}^{7} (\text{variable rank} \times \text{weight of criterion})$$

The DRASTIC method assumes that the flow of the groundwater is linear. This assumption is fully acceptable for the porous media in the aquifer of the central Vistula River valley. Final evaluation of vulnerability depends on the precision and accuracy of hydrological character, which is very high for the KNP region. The DRASTIC evaluation produces a map that shows the distribution of values of the vulnerability index IPZ_Σ (Figure 2).

Table 1.   Rating and weight of criterion for DRASTIC criterions with assigned weights (after Aller et al. 1987 – modified for Vistula River valley).

No.	Criterion	Classes of criterion	Weight of criterion	Rank
1	Depth to groundwater water table [m]	>5 m	5	7
		3.1–5 m		8
		1.1–3 m		9
		<1 m		10
2	Net recharge [mm/year]	50–75	4	2
		101–130		3
		151–180		4
		181–250		5
		>250		6
3	Lithology of aquifer	Sandy clay, loam, loam and sands	3	2
		Sandy loam, sands		3
		Sands, sandy loam		4
		Sands		6
		Sands, gravel		8
4	Soil media	Loam	2	5
		Sandy loam		6
		Shrinking clay		7
		Peat		8
		Thin anthropogenic		9
		Absent		10
5	Topography (slope) [%]	2.9–3.9	1	7.5
		2.5–2.9		8
		2.0–2.5		8.5
		1.6–2.0		9
		1.0–1.6		9.5
		1.0–0.0		10
6	Lithology of unsaturated zone	Clay	5	2
		Silty loam		3
		Loam		4
		Sands		6
		Sands, gravel		8
7	Hydraulic conductivity of aquifer [m/day]	<4	3	1
		4–12		2
		13–28		4
		29–40		6
		41–80		8

Assignments of each criterion (hydrogeological data, soil assignments, and topography) were plotted on several maps (scale 1:50,000) with the use of vectoring, calculations and visualisations in the following computer programs: ARC/INFO 8.0.1, ArcView 3.2, AVSpatialAnalist 1.1, AVArcPress 2.0 produced by ESRI, operating on Sun Solaris and Win NT platforms. Modeling called for all criteria to be brought into a form of pseudo-continuous distributions, expressed in a form of nets of natural mesh with a resolution of 100 m × 100 m (block 100 m × 100 m, more than 65,000 blocks).

Basing on the initial data and calculations conformable with DRASTIC methods slightly modified but conforming to the specifics of the study area, classification of groundwater

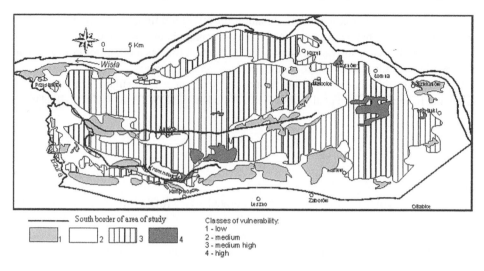

	South border of area of study	Classes of vulnerability:
	1, 2, 3, 4	1 - low 2 - medium 3 - medium high 4 - high

Figure 2. Groundwater vulnerability map of part of Vistula River valley, Kampinoski National Park area based on the DRASTIC method.

Table 2. Distribution of the vulnerability classes – DRASTIC model.

IPZ$_\Sigma$	Classes of the relative vulnerability	Area [km^2]	Percentage share of classes
<100	Very low	0.01	0.002
100–125	Low	52.82	8.63
126–150	Medium	228.05	37.25
151–175	Medium high	318.86	52.08
176–200	High	12.53	2.05
>200	Very high	0.03	0.005
Σ	612,3 km^2 – area of hydrogeological unit		100%

vulnerability was obtained, dependent on the range of the IPZ index. The following types of groundwater vulnerability were assigned: very low (IPZ$_\Sigma$ < 100), low (IPZ$_\Sigma$ from 100 to 125), medium (IPZ$_\Sigma$ from 126 to 150), moderately high (IPZ$_\Sigma$ from 151 to 175), high (IPZ$_\Sigma$ from 176 to 200), very high (IPZ$_\Sigma$ > 200).

The middle Vistula River valley is characterized by moderate (37% of the area or 228 km^2) and moderately high (52% of the area or over 318 km^2) vulnerability to contamination. In the study area about 12.5 km^2 is an area of high vulnerability. Remaining vulnerability classes are located on considerably smaller areas (Figure 2, Table 2).

3.2 *Migration time of conservative contamination as a criterion of vulnerability*

Infiltration is the vertical water movement downward through the unsaturated zone. Estimation of infiltration time is a key factor in determining the vulnerability of groundwater. The migration time of conservative contamination (infiltration time through the unsaturated zone) is one of rational criteria used in evaluating groundwater vulnerability for contamination. Approximate determination for migration time can be achieved based on time of water exchange in a rock formation assuming piston-flow model.

Three models were constructed employing infiltration time as a criterion for evaluation of groundwater vulnerability to contamination.

Model 1 – infiltration time through the unsaturated zone was calculated with the use of following formula (Wosten et al. 1986; Haith & Laden 1989; Witczak & Żurek 1994):

$$t_a = \sum_{i=1}^{n} \frac{m_i \cdot (w_o)_i}{I_e} \tag{1}$$

where:

 m_i – thickness of successive layers of unsaturated zone profile [m]

 w_o – average volumetric moisture of successive layers of unsaturated zone [–]

 I_e – infiltration of atmospheric precipitation deep into the soil profile [m^3/m^2 × year] obtained through multiplication of infiltration rate (w_i [%]) by the volume of precipitation.

Infiltration time according to formula [2] directly depends on volumetric moisture of sediments, most often assumed data tables compilations (Wiłun 1987).

Model 2 – Infiltration time calculated according to Bindeman's formula:

$$t_a = \sum_{i=1}^{n} \frac{m_i n_o}{\sqrt[3]{I_e^2 k'}} \tag{2}$$

where:

 n_0 – effective porosity [1]

 k' – vertical hydraulic conductivity of unsaturated zone [LT^{-1}] rest of denotations it as in formula [1]

The Bindeman equation states that infiltration time, excluding thickness of unsaturated zone which is taken into account in all formulas, primarily depends on infiltration intensity and effective porosity but in lesser importance on infiltration coefficient.

Model 3 – to evaluate infiltration time, Macioszczyk (1992) proposed a modified formula:

$$t_a = \sum_{i=1}^{n} \frac{m_i (w_o)_i}{\sqrt[3]{I_e^2 k'}} \tag{3}$$

Denotations same as in formulas [1], [2].

Migration time of conservative contamination through unsaturated zone ($t = t_a + t_p$) is a sum of infiltration through unsaturated zone (t_a) and eventual percolation through deeper cover (t_p). This is the reason why separate calculations were performed for the soil layer cover (average thickness in the studied area = 35 cm) and for the remaining unsaturated zone thickness. The result is the sum of flow time through both zones.

The calculations indicate variable migration times for conservative contamination depending on the model chosen (Table 3).

Table 3.   Migration time of conservative contamination in Vistula River valley (area of the Kampinoski National Park), percentage share of classes.

Vulnerability classes	Migration time of conservative contamination [year]	Percentage share of classes		
		Model 1	Model 2	Model 3
Very high	<0.083 (30 days)[1)]	0.4	5.4	0,0
	0.083–0.5	4.2	48.7	55.3
High	0.5–1	10	33.7	28.2
	1–3	64.7	12.2	12.1
Medium high	3–5	8.1	0	2.0
	5–10	2.6	0	1.9
Medium	10–15	7.9	0	0.5
	15–20	0.3	0	0
Low	20–25	1.0	0	0
	25–30	0.2	0	0
Very low	>30	0.6	0	0

[1)] according to Polish Low (till 2001 year) – range of intake protection zone.

Using the formula in Model 1, which is highly recommended in hydrological calculations conducted in Poland, a surprisingly long infiltration time is indicated. In the area were the groundwater table not exceeds 1.5 m below terrain surface and the unsaturated zone is mostly sandy, the infiltration time is from 30 days to 15 years.

Results obtained by using the most commended formula (Model 2) in Polish hydrogeological studies are expected lower as the formula does not take into account the volumetric moisture of the unsaturated zone, just the porosity. Almost the entire study area has an infiltration time of 30 days to 1-year for conservative contamination from the terrain surface.

Infiltration time, determined by formula [3] is from 30 days to 10 years, but almost 96% of the study area is characterized by an infiltration time of 30 days to 3 years.

4   CONCLUSIONS

DRASTIC method enabled creation of general classification and map of natural vulnerability for the study area, based on calculated values of the IPZ index. This may be useful tool for environmental management. For areas where groundwater table in porous medium is shallow – river valley, and not isolated by low permeability natural from the terrain surface, for example in river valleys, an additional method to use is the migration time of conservative contamination from the terrain surface. Supplementing the base DRASTIC model can be conducted for an entire study area of study or selected parts, characterized by high natural vulnerability or perhaps in cases where there are planned changes in development. The resulting maps, obtained from U.S. EPA DRASTIC, of groundwater vulnerability show that in river valley – area of study is generally characterized by medium and medium high natural vulnerability. Infiltration time, determined by different formulas is generally from 30 days to 3 years.

Calculations performed with different methods, in this case, utilizing different parameters (in consequence different maps) can become basis of scenario maps (Żurek et al.

1994) showing varying concepts for protection of groundwater depending on terrain development.

ACKNOWLEDGEMENTS

The author would like to acknowledge the State Committee for Science Research in Poland for providing financial support (project 8T12B00721), invaluable help and support of Kampinoski National Park, Geological Bureau Geonafta Warsaw, Poland.

REFERENCES

Aller, L., Bennett, T., Lehr, J.H., Petty, R.J. & Hackett, G. 1987. DRASTIC: a standardized system for evaluating ground water pollution potential using hydrogeological settings. US-EPA Reports, 600/2-87-035, Washington DC, USA.

Centre Naturopa 1998. The paneuropean ecological network. *Questions and answers*. 4.

Duijvenbooden, W. & Waegeningh, H. G. 1987. Vulnerability of Soil and Groundwater to Pollutants, *Proceedings and Information. No. 38 the International Conference held in Netherlands, 1987*, TNO Committee on Hydrological Research, Delft, The Netherlands.

Haith, D.A. & Laden, E.M. 1986. Screening of groundwater contaminations by travel-time distributions. *Journal of Environmental Engineering*, ASCE vol. 115, 3: 497–512.

Krogulec, E. 1997a. Numeryczna analiza struktury strumienia filtracji w strefie krawędziowej poziomu błońskiego Wyd. Uniwersytetu Warszawskiego. Warszawa.

Krogulec, E. 1997b. Identyfikacja wartości wskaźnika filtracji na modelach numerycznych strefy krawędziowej poziomu błońskiego. In: *Modelowanie matematyczne w hydrogeologii i ochronie środowiska*. Częstochowa XI: 93–97.

Krogulec, E. 2001a. Monitoring shallow groundwater within protected areas – an example of local "groundwater monitoring network in Kampinos National Park (Poland). *Proc of Confer. Hydrogeochemia 2001*", Bratislava: 117–123.

Krogulec, E. 2001b. The role of Blonski Level in groundwater recharge of Kampinos National Park (Poland). Conference *"New Approaches Characterizing Groundwater Flow"* (Eds. K. Seiler, S. Wohnlich). A.A. Balkema Publ.: 555–558.

Krogulec, E. 2003. Hydrogeological conditions of the Kampinos National Park (KNP) region. *Ecohydrology and Hydrobiology*, vol. 3, 3: 257–266.

Limisiewicz, P. 1998. Ocena podatności wód podziemnych na zanieczyszczenie w wybranych zlewniach Dolnego Śląska, Rozprawa doktorska, Uniwersytet Wrocławski, Instytut Nauk Geologicznych, Zakład Hydrogeologii, Wrocław.

Macioszczyk, T. 1992. Parametry hydrogeologiczne. In: *W służbie polskiej hydrogeologii*. Wyd. AGH, Kraków: 191–196.

Vrba, J. & Zaporozec, A. (eds.), 1994. Guidebook on mapping groundwater vulnerability, IAH, *International Contributions to Hydrogeology*, vol. 16 Heise Verlag, Hannover.

Wiłun, Z. 1997. Zarys geotechniki. Wyd. Kom. i Łączności, Warszawa.

Witczak, S. & Żurek, A.1994. Wykorzystanie map glebowo-rolniczych w ocenie ochronnej roli gleb dla wód podziemnych. In: *Metodyczne podstawy ochrony wód podziemnych* (eds. A.S. Kleczkowski). Projekt badawczy KBN. Wyd. AGH, Krakow.

Wosten, J.H., Bannink, J.J., De Gruijter, J. & Bouma, J. 1986. A procedure of identify different groups of hydraulic conductivity and moisture retention curves for soils horizons. *Journal of Hydrology*, v. 86: 133–145.

Żurek, A. Kleczkowski, A.S. & Witczak, S. 1994. Mapa scenariuszowa ochrony wód podziemnych na przykładzie zbiornika Opole-Zawadzkie (GZWP 333). In: *Metodyczne podstawy ochrony wód podziemnych* (eds. A.S. Kleczkowski). Grant KBN. Wyd. AGH, Kraków.

CHAPTER 12

Development of an integrated methodology for the assessment of groundwater contamination by pesticides at the catchment scale

B. Leterme,[1] M.D.A. Rounsevell,[1] D. Pinte,[2] J.D. Piñeros-Garcet[2] & M. Vanclooster[2]

[1] *Department of Geography, Université Catholique de Louvain, Louvain-la-Neuve, Belgium*
[2] *Department of Environmental Sciences and Land Use Planning, Unité Génie Rural, Université Catholique de Louvain, Louvain-la-Neuve, Belgium*

ABSTRACT: A methodology to assess groundwater contamination by pesticides is presented and is applied to the Dyle catchment area. The development of the methodology follows the following sequence in the analysis of the problem. (1) Groundwater vulnerability is first approached *via* the monitored data of pesticide concentrations in the aquifer. The limitations of this approach are discussed; (2) then the methodology evolves towards the use of pesticide fate modelling. Other factors affecting the vulnerability assessment, such as data availability are presented and discussed. (3) Different modelling techniques are adopted to comply with these constraints (4) finally, the methodology lead to the use of conditional simulations at the catchment scale to assess groundwater vulnerability. The vulnerability assessment used conditional simulations to perform the spatialization of the point information derived from the leaching simulations on soil profiles. This research suggests that better results can be expected when the methodology is progressively adapted to the findings made during the study, rather than sticking to a pre-defined schedule.

1 INTRODUCTION

The protection of groundwater is receiving increasing attention, as exemplified in Europe by the recent Water Framework Directive (EU 2000). Groundwater quality issues deal with the presence and persistence of pollutants in groundwater. Pollution problems cover a wide range of processes, depending on the substance (heavy metals, pesticides, nitrate...), the pollution type (diffuse *vs.* point sources), or the aquifer type (porous/karstic, free/confined...). The vulnerability concept is useful to study these pollution issues. It is proposed to start here from the definition of groundwater vulnerability given by Duijvenbooden and Waegeningh (1987): for quality issues, groundwater vulnerability is defined as *"the sensitivity of groundwater quality to an imposed contaminant load, which is determined by the intrinsic characteristics of the aquifer"*. In this study the discussion will be limited to the quality aspect of groundwater vulnerability, while quantity aspects are not treated. While vulnerability should not be mixed up with pollution risk (which depends not only on vulnerability but also on the existence of pollution loading entering the subsurface environment), it is nevertheless appropriate to pick out here that groundwater vulnerability assessments have to consider the nature of the pollutant. Indeed, effective vulnerability maps have to be specific to a substance or a substance class, because a number of important processes depend on the pollutant properties. This is illustrated by Worrall et al. (2002), who stressed that interaction between site and chemical factors represented the most important control on the occurrence of pesticides in groundwater. A strictly intrinsic vulnerability assessment would then neglect important processes and produce biased results. So, the definition above could be completed with "*...and their interactions with the contaminant*".

Groundwater vulnerability assessment methods are usually classified into three main categories: overlay and index methods, methods employing process-based models, and statistical methods (NRC 1993). The first category, overlay and index methods, is based on the combination of various parameters into a score or an index, possibly using weights to reflect the relative importance of different factors. An example of such a method is the DRASTIC index (Aller et al. 1985), and typical variables used include depth to groundwater, organic matter content, net recharge, etc. In the second category, vulnerability assessment methods can have a range of complexity depending on the use made of process-based models. These models may be coupled to other models (e.g. two- or three-dimensional models of the saturated zone) integrating different parts of the process under study (see e.g. Fogg et al. 1999; Vanclooster et al. 2002), or they may simply be used to derive an index or ranking from simulation results (e.g. Bleecker et al. 1995). Statistical methods are basically developed using data of monitored areal distribution of a groundwater contaminant. Approaches such as discriminant or cluster analysis have been used to describe relationships between groundwater vulnerability and site properties (e.g. Troiano et al. 1999). The final output of statistical methods may be for instance the probability, possibly conditioned by other variables, of a substance to be detected at a given location.

It might be interesting to discover the fundamental reasons behind the choice of the authors cited above (Fogg et al. 1999; Bleecker et al. 1995; Troiano et al. 1999) for one or another vulnerability assessment method. It is likely that sometimes the choice is dictated by the author's personal preferences: e.g. after the development of a process-based leaching model, the modeller could be inclined to test and use his/her model to perform vulnerability assessments. However, the selection of a method should be as objective as possible. This choice should reflect the data availability and the level of confidence of the available tools (models, indices, geostatistical methods, etc.).

The present paper follows the different steps of a reasoning process aiming at the assessment of groundwater vulnerability to atrazine at the catchment scale. Given data availability or other factors (e.g. dominant processes, scale consideration), different possibilities are explored and allow the simultaneous development of an effective vulnerability assessment approach.

2   STUDY AREA

The present research was conducted at the catchment scale, using the Dyle catchment area ($580\,km^2$; Figure 1a) as a case study. The Dyle river is located in the central part of Belgium, in a loamy region of intensive cropping (mainly wheat, sugar beets, maize and barley). From a hydrogeologic standpoint, the catchment is entirely connected to the same

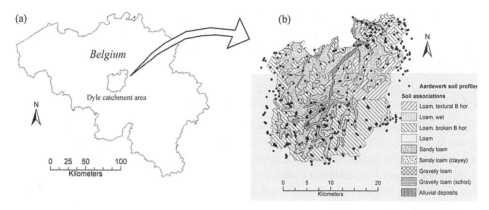

Figure 1.   (a) Location of the study area. (b) Digital soil information available: soil associations map (Maréchal & Tavernier 1974) and Aardewerk profiles (Orshoven & Vandenbroucke 1993).

underlying groundwater body, known as the Brusselian aquifer. It is an unconfined aquifer located in tertiary sands, and overlain by a quaternary loess layer of variable thickness (0 to 15 m). In the North-Eastern part of the catchment, the Brusselian sands are progressively replaced by an underlying body of secondary chalk. The Brusselian sands are outcropping in the valleys where sand and sandy loam occur.

3   DATA

In the catchment, 41 drinking water wells and 6 galleries extract water from the aquifer for different drinking water companies. Regional authorization for drinking water exploitation is subjected to the analysis of the presence of plant protection products at least once a year. Among all available pesticide data, the present research focused on atrazine as this is one of the most problematic pesticides in the Brusselian aquifer (Belgaqua – Phytofar 2002; DGRNE 2003), and it is the pesticide for which most data are available in the Dyle catchment (sampling period from the early nineties onwards). In the study area, atrazine is used on silage maize, but it is also used in the non-agricultural weed control programmes.

Soils data (Figure 1b) originate from two different data sets. The first is the Belgian soil associations map at a scale of 1:500,000 (Maréchal and Tavernier 1974), and the second is the Aardewerk soil profile database (Orshoven and Vandenbroucke 1993) which consists of more than 10,000 soil profiles covering Belgium, among which 300 are located in the arable part of the study area.

4   GROUNDWATER VULNERABILITY ASSESSMENT

4.1   *Outline of the integrated method to assess groundwater vulnerability*

The proposed integrated methodology is based on 5 steps, which represent a logical sequence in the analysis of the problem. The steps are (a) vulnerability assessment directly based on monitored data, (b) vulnerability assessment based on the interpolation of monitored data,

(c) vulnerability assessment using a leaching model and based on the soil association map and representative soil profiles, (d) vulnerability assessment using a leaching model and based on data from all soil profiles, (e) vulnerability assessment using a conditional simulation of the 90th percentile of yearly averaged atrazine concentrations.

4.2   *Vulnerability assessment directly based on monitored data*

Atrazine concentrations observed in groundwater constitute the most immediate indication of groundwater vulnerability. Rather than applying various types of models and methods which sometimes involve large uncertainties, monitored data directly reflect groundwater quality with an uncertainty only resulting from measurement and location errors, which are usually low. Despite the point nature of monitored data, Worrall (2002) used them in the study of Californian groundwater, to directly assess groundwater vulnerability *via* the definition of well catchments, i.e. the catchment areas of each well or gallery. Following this spatialization, Bayesian methods were used to calculate a continuous measure of well (catchment) vulnerability compared to the entire study area. However, this approach is not applicable to the present research, due to the relatively poor availability of monitored data. Indeed, 404 observations (164 of which belonged to 4 wells) were available for the Dyle catchment; while in the study of Californian groundwater Worrall (2002) had almost 2000 observations of atrazine presence or absence, which ensured statistical consistency in the analysis.

Thus, due to the limited size of the monitored data set, a vulnerability assessment method directly based on groundwater observations is not applicable here. The information content of monitored data is nevertheless significant, and it may be used in further developments as discussed later.

4.3   *Vulnerability assessment based on the interpolation of monitored data*

Whilst monitored data are not sufficiently numerous in the present case to directly calculate probabilities of vulnerability using Bayesian methods, it might be relevant to perform spatial and temporal interpolations between these data, in order to capture the vulnerability of the Dyle catchment area. Spatial and temporal variability of atrazine concentrations in groundwater was therefore explored using semi-variogram analyses, with the BMELib package (Christakos et al. 2002). After drawing the first mean temporal semi-variogram, it was soon realized that monitored data had to be split into two subsets: wells reflecting pollution processes from non-agricultural sources *vs.* wells subject to diffuse pollution from agricultural sources or showing a significant temporal trend and hence inducing non-stationarity. When this was done, the temporal analysis on the detrended "agricultural" data set allowed an exponential model (with a nugget effect) to be fitted with a *range* of approximately 300 days. The mean spatial semi-variogram, with the *range* fitted around 550 m (cf. dashed line), is presented in Figure 2. The gray thin lines display several experimental semi-variograms obtained with small changes in classes definition. From these, an average semi-variogram was calculated (white dots), to which an exponential model was fitted (black thick line). Assuming that the aquifer is homogeneous, this semi-variogram could be used to interpolate atrazine concentrations between monitoring stations. However, the spatial distribution of "agricultural" wells and the spatial extent to which this interpolation could be done with a relatively high confidence do not justify an application to the agricultural part of the Dyle catchment.

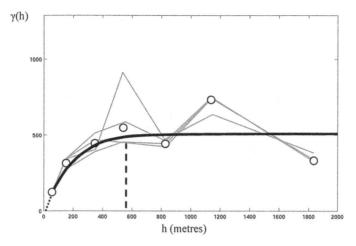

Figure 2.   Semi-variogram of the spatial variability of atrazine concentrations in the Brusselian aquifer.

Valuable indications about the spatio-temporal persistence of atrazine in groundwater were collected here. Yet, the results suggest that monitored atrazine concentrations in groundwater alone cannot be used as a basis for the vulnerability assessment at the scale of the catchment area, even though this information may be useful in an integrated vulnerability assessment, as discussed later.

4.4   *Vulnerability assessment using a leaching model and based on the soil association map and the soil profiles*

Given the above conclusions, the next step consisted in approaching the vulnerability assessment *via* the simulation of atrazine fate from the surface. It was decided to use a pesticide leaching model rather than an index, because it was thought that the available soils data (e.g. detailed horizons description for the Aardewerk profiles) could be better exploited by a process-based model, for example by considering depth dependent processes. As the objective of the modelling study was to obtain a spatially-distributed vulnerability assessment, the model to be used has to allow easy GIS implementation at the catchment scale. Besides, the criteria that should be considered when selecting a model for a simulation exercise are: (i) a reasonably validated model; (ii) a verifiable model code; (iii) the representation of key transport processes at the scale for which the model will be applied; (iv) reasonable modelling assumptions; and (v) availability of input data for the given model (FOCUS 2000; Mulla & Addiscott 1999). The model GeoPEARL (Tiktak et al. 2002) was finally chosen to perform the simulations. The GeoPEARL code results from the coupling of the widely validated 1-D leaching model PEARL (Tiktak et al. 1996) with spatial input files that can be derived from a standard GIS. The spatial resolution was chosen as 100×100 m, which constitutes a compromise between accuracy and the resolution of input data (soil map, land use).

An overview of the validation status and limitations of PEARL model as compared to other reference pesticide leaching models have been given by Vanclooster et al. (2000) and

Trevisan et al. (2003). A major limitation of the current version of the model is the absence of preferential flow simulation, though recent modifications of the code to remediate to this have been proposed (Jarvis et al. 2003). Other processes that cannot be simulated by GeoPEARL include transport in the vadose zone below the soil profile and in the groundwater body. Different techniques can be used to fill this last gap (e.g. an artificial extent of the soil profile, using simplified transfer functions for simulating transport in the vadoze zone) but were not considered in the present analysis. As a first approach, however, it was decided to focus on atrazine leaching at the bottom of the soil profile, i.e. at 2 m depth. Yearly (spring) applications of atrazine were simulated for silage maize cropping. The simulation period was fixed as 1980 to 2002 to match the period covered by monitored data, with a buffer for the initialization years.

Early simulations showed that soil parametrization, and especially organic matter, was by far the most sensitive factor affecting atrazine leaching. Conversely, the variability of climatic factors had a lower importance. This is logical because at the scale of the Dyle catchment, spatial variability in the weather stations is small, unlike studies at the regional scale where this factor was shown to be of primary importance (e.g. Bleecker et al. 1995). It was decided therefore to concentrate on the development of a method which efficiently uses the available soil information in the study area.

The initial approach was to combine the information for both the soil associations map and the Aardewerk soil profiles. For each soil association, only the profiles sampled on arable land were selected. As the hydrological processes simulated in GeoPEARL are non-linear, the simulations were first run on the individual soil profiles and the results aggregated to the soil association level, in order not to underestimate leaching. Figure 3 shows

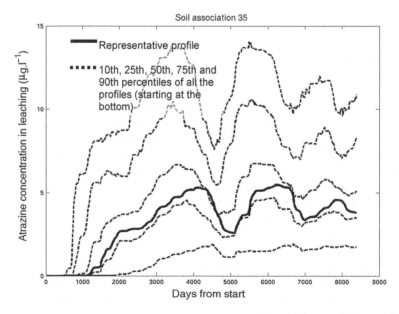

Figure 3.   Atrazine concentration in leachate at 2 m depth simulated by GeoPEARL, for the representative profile of a soil association (dashed line) and percentiles for all the Aardewerk profiles of that association (solid lines).

indeed that, for a given soil association, leaching at the bottom of a representative profile (calculated by a weighted average of the horizon properties of the selected individual profiles) is underestimated compared to the median of the leaching of all the individual soil profiles. This is certainly a consequence of the model non-linearity (Rastetter et al. 1992).

A vulnerability map could then be designed by taking e.g. the 90th percentile of the leaching simulated for the set of soil profiles within each soil association. However, this method would not produce a useful spatial assessment of the vulnerability since a single value would be allocated to each soil association, while most of the agricultural part of the catchment is covered by only three different soil associations (loam, textural B horizon; loam, broken B horizon; and sandy loam; cf. Figure 1b).

Moreover, the analysis of the simulation results revealed that the information given in the soil association map was not appropriate for the scale at which groundwater vulnerability is assessed. For example, for the main soil properties (texture, organic matter content, horizons sequence) it was observed that there is a higher variability among the profiles of a given soil association than between different soil associations, and this variability is further reflected in the simulated atrazine leaching. Soil maps at the 1:20,000 scale, currently available only on paper, would perhaps turn out to be more useful and bring effective additional information if they were made available in a digitized form in the future.

4.5 *Vulnerability assessment using a leaching model and based on the soil profiles*

As the information contained in the 1:500,000 soil associations map is inappropriate in the present study, the arable part of the catchment area on which vulnerability has to be assessed can now be considered as one single unit combining the totality of the arable Aardewerk profiles. Here, it was necessary to determine an indicator of the vulnerability of a soil profile based on the simulation results. It was decided to take the 90th percentile of all the yearly averaged atrazine concentrations for the leaching at 2 m depth. This percentile is often referred to as a realistic worst case scenario (FOCUS 2000). Here, the spatialization of this point information was still an issue. A semi-variogram analysis was therefore performed to investigate the spatial variability of this variable on which the vulnerability assessment will be based. The experimental and fitted semi-variograms are displayed on Figure 4, showing a clear spatial dependency despite a non-negligable nugget effect. The *range* is situated at about 6,000 m, a value that relates to the strong influence of the spatial variability of soil organic matter for which a *range* was evaluated at about 4,000 m (not shown here). It is possible in applying kriging techniques to obtain interpolated values between the soil profiles. However, while kriging is a powerful technique when a dense network of points is available, the distribution of Aardewerk profiles in the Dyle catchment leaves areas where the linear interpolation of kriging would give interpolated values with a high uncertainty. Moreover conditional simulations allow an easier detection of areas susceptible to be highly vulnerable to groundwater contamination. This reasoning lead to the last step of the development of the vulnerability assessment method.

4.6 *Vulnerability assessment using conditional simulations of the 90th percentile of yearly averaged atrazine concentrations*

The clarity of the semi-variogram (Figure 4) suggests that the spatial correlation of the 90th percentiles should be considered to construct a vulnerability map. In this respect, the

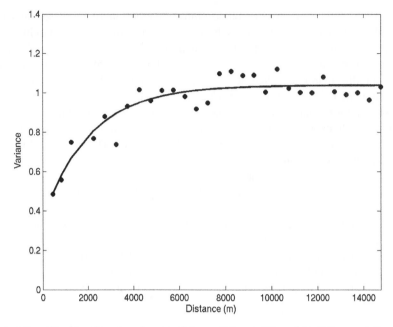

Figure 4.   Omnidirectional semi-variogram of the spatial variability of the 90th percentile of yearly averaged atrazine concentration, simulated at the bottom of the soil profiles.

application of a conditional spatial simulation seems relevant here, as this will respect both the point values simulated by GeoPEARL and the spatial dependency of the variable. 50 simulations were performed to capture the spatial pattern that is the most likely to occur given the conditionalization on the Aardewerk based PEARL simulations. A vulnerability map can be derived from these 50 conditional simulations, taking e.g. the mean or a certain percentile value in every location. The 90th percentile of all the simulations was chosen here as the indicator of vulnerability to atrazine contamination (Figure 5). So, Figure 5 displays the 90th percentile of all the 50 simulated values of this indicator (the 90th percentile of yearly averaged atrazine concentrations). The conditionalization values of the soil profiles are also displayed on Figure 5. It can be observed that vulnerability is often higher near highly vulnerable soil profiles, but not systematically when less vulnerable soil profiles are close. Conversely, homogeneous areas of low vulnerability are found when a significant number of neighbours soil profiles have a low vulnerability (e.g. in the south-west of the catchment).

5   DISCUSSION

The vulnerability assessment adopted here results from the adaptation of the methodology to a set of constraints. These constraints were mainly driven by data availability (quantity or quality; e.g. limited number of monitored data or inappropriate scale of the soil association map), but also by the nature of the processes studied (e.g. spatial correlation *range* of different variables) or by the tools availability (e.g. choice of the leaching model). This clearly illustrates that a high number of factors can have an influence on the selection of a vulnerability assessment method. The latter will therefore be probably different for

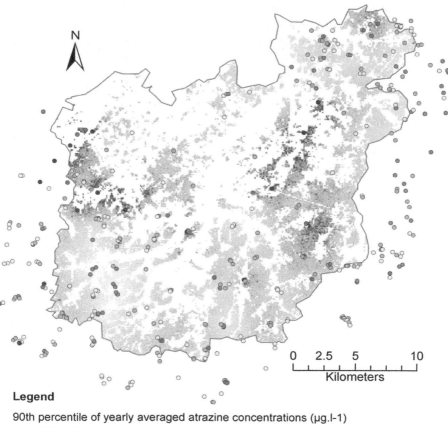

Legend

90th percentile of yearly averaged atrazine concentrations (µg.l-1)

- 0.03 - 2.54
- 2.55 - 4.87
- 4.88 - 7.19
- 7.20 - 9.81
- 9.82 - 13.13
- 13.14 - 19.81
- 19.82 - 31.59
- 31.60 - 51.70

90th percentile of all the simulations (µg.l-1)

46.85

3.12

White: non-arable areas

Figure 5.   Vulnerability assessment based on the 90th percentile of the conditional simulation values. The variable simulated is the 90th percentile of yearly averages atrazine concentrations ($\mu g.1^{-1}$) calculated on the soil profiles (dots on the figure).

different case studies if it is aimed to fully exploit the available information. However, taking into account the factors cited above in a scientifically objective way can serve as a guideline for the development of a groundwater vulnerability assessment in any situation.

The vulnerability assessment method that was finally adopted in this paper concerns atrazine leaching at the bottom of a 2 m soil profile. These results are therefore only a part of an integrated assessment that should consider the vadose zone but also the groundwater body itself. Indeed, as stressed by Fogg et al. (1999), the potential for attenuation of contaminant concentrations by both physical and chemical processes is substantial between the water table and the well intakes. In this respect, different methods, such as transfer functions, are available to account for the contaminant transport through the vadose zone (Javaux 2004). This point will be investigated in the future. As far as the groundwater body is concerned, a 3-D model of the Brusselian aquifer might become available in the near future. This would certainly be very useful to integrate that part of the environment in the vulnerability assessment, and the information contained in monitored data could thereby be exploited. Another planned development of the methodology concerns the inclusion of non-agricultural sources of contamination in the vulnerability assessment.

6   CONCLUSIONS

This paper presented the development of an integrated vulnerability assessment method, using the groundwater contamination by atrazine in the Dyle catchment area as a case study. The research followed a logical sequence in the analysis of the problem. The first steps investigated the possibility of using monitored data of atrazine concentrations in groundwater as primary information for the vulnerability assessment. Despite the valuable information content of these data, inadequacies forced the methodology to be oriented towards pesticide fate modelling. The methodology eventually lead to the use of conditional simulations to perform the spatialization of the point vulnerability information derived from the leaching simulations on the soil profiles.

ACKNOWLEDGEMENT

This work was supported by the *Fonds National de la Recherche Scientifique* (FNRS, Belgium). The authors are grateful to the *Direction générale des Ressources naturelles et de l'Environnement* (DGRNE) and to the *Vlaamse Maatschappij voor Watervoorziening* (VMW) for delivering the data on groundwater quality.

REFERENCES

Aller, L., Bennett T., Lehr, J.H. & Petty, R.J. 1985. DRASTIC: a standardized system for evaluating ground water pollution potential using hydrogeologic settings. US.-EPA/600/2–85/018, Ada Oklahoma, USA.

Belgaqua – Phytofar 2002. Livre Vert: p. 41.

Bleecker, M., DeGloria, S.D., Hutson, J.L., Bryant, R.B. & Wagenet, R.J. 1995. Mapping atrazine leaching potential with integrated environmental databases and simulation models. *Journal of Soil and Water Conservation* 50(4): 388–394.

Christakos, G., Bogaert, P. & Serre, M.L. 2002. Temporal GIS. Advanced Functions for Field-Based Applications. Springer-Verlag, New York, NY.

DGRNE 2003. Tableau de bord de l'environnement wallon 2003. Ministère de la Région wallonne, Direction Générale des Ressources Naturelles et de l'Environnement: p. 143.

Duijvenbooden, W. van & Waegeningh, H.G. van 1987. Vulnerability of soil and groundwater to pollutants. *Proceedings and Information No. 38 of the International Conference held in the Netherlands, in 1987*, YNO Committee on Hydrological Research, Delft, The Netherlands.

EU 2000. Directive 2000/60/EC of the European Parliament and of the Council of 23 October 2000 establishing a framework for Community action in the field of water policy. Official Journal of European Communities (L 327): 1–72.

FOCUS 2000. FOCUS groundwater scenarios in the EU review of active substances. Report of the FOCUS Groundwater Scenarios Workgroup, EC Document Reference Sanco/321/2000 rev.2: p. 202.

Fogg, G.E., LaBolle, E.M. & Weissmann, G.S. 1999. Groundwater Vulnerability Assessment: Hydrogeologic Perspective and Example from Salinas Valley, California. In: *Assessment of Non-Point Source Pollution in the Vadose Zone* (Corwin, Loague & Ellsworth eds.), Washington, American Geophysical Union: 45–61.

Jarvis, N., Boesten, J.J.T.I., Hendriks, R., Klein, M., Larsbo, M., Roulier, S., Stenemo, F. & Tiktak, A. 2003. Incorporating macropore flow into FOCUS PEC models. In: *XII Pesticide chemistry symposium "Pesticide in air, plant soil and water system"* (Del Re, Trevisan & Capri eds.), Piacenza, Italy 4–6 June 2003.

Javaux, M. 2004. Solute transport in a heterogeneous unsaturated subsoil. PhD Dissertation, Faculté d'ingénierie biologique, agronomique et environnementale. Louvain-la-Neuve, Université catholique de Louvain: p. 194.

Maréchal, R. & Tavernier, R. 1974. Atlas de Belgique. Pédologie. Commentaire des planches 11 A et 11 B. Extraits de la carte des sols – Carte des Associations de sols. Comm. Nat. Atlas, Gent.

Mulla, D.J. & Addiscott, T.M. 1999. Validation Approaches for Field-, Basin-, and Regional-Scale Water Quality Models. In: *Assessment of Non-Point Source Pollution in the Vadose Zone* (Corwin, Loague & Ellsworth eds.), Washington, American Geophysical Union: 63–78.

National Research Council 1993. Ground Water Vulnerability Assessment. Washington D.C., National Academy Press: p. 204.

Orshoven, J. van & Vandenbroucke, D. 1993. Guide de l'utilisateur de AARDEWERK. Base de données de profils pédologiques. IRSIA – COBIS – Instituut voor Land-en Waterbeheer, Katholieke Universiteit Leuven: p. 46.

Rastetter, E.B., King, A.W., Cosby, B.J., Hornberger, G.M., O'Neill, R.V. & Hobbie, J.E. 1992. Aggregating Fine-Scale Ecological Knowledge to Model Coarser-Scale Attributes of Ecosystems. *Ecological Applications* 2(1): 55–70.

Tiktak, A., De Nie, D.S., van der Linden, A.M.A. & Kruijne, R., 2002. Modelling the Leaching and Drainage of Pesticides in the Netherlands: The GeoPEARL model. Agronomie 22: 373–387.

Tiktak, A., van der Linden, A.M.A. & Merkelbach, R.C.M. 1996. Modelling pesticide leaching at a regional scale in the Netherlands. RIVM report n° 715801008. Bilthoven, National Institute of Public Health and the Environment: p. 75.

Trevisan, M., Padovani, L., Jarvis, N., Roulier, S., Bouraoui, F., Klein, M. & Boesten, J.J.T.I. 2003. Validation status of the present PEC groundwater models. In: *Proc. XII Pesticide chemistry symposium. Pesticide in air, plant soil and water system*. Piacenza (Del Re, Trevisan and Capri eds.), Italy 4–6 June 2003.

Troiano, J., Marade, J. & Spurlock, F. 1999. Empirical Modeling of Spatial Vulnerability Applied to a Norflurazon Retrospective Well Study in California. *Journal of Environmental Quality* 28: 397–403.

Vanclooster, M., Boesten, J.J.T.I., Trevisan, M., Brown, C., Capri, E., Eklo, O.M., Gottesbüren, B., Gouy, V. & van der Linden, A.M.A. 2000. A European test of pesticide-leaching models: methodology and major recommendations. *Agricultural Water Management* 44: 1–21.

Vanclooster, M., Javaux, M., Hupet, F., Lambot, S., Rochdi, A., Piñeros-Garcet, J.D. & Bielders, C. 2002. Effective approaches for modelling chemical transport in soils supporting soil management at the larger scale. In: *Sustainable land management – environmental protection. A soil physical approach.* (Pagliai and Jones eds.). Advances in geo-ecology 35: 171–184.

Worrall, F. 2002. Direct assessment of groundwater vulnerability from borehole observations. In: *Sustainable Groundwater Development.* (Hiscock, Rivett & Davison eds.), The Geological Society of London, London, 193: 245–254.

Worrall, F., Besien, T. & Kolpin, D.W. 2002. Groundwater vulnerability: interactions of chemical and site properties. *The Science of the Total Environment* 299: 131–143.

CHAPTER 13

Hydrological controls of the groundwater vulnerability maps (case study of the lower Vistula valley near Plock, Poland)

A. Magnuszewski

Warsaw University, Faculty of Geography and Regional Studies, Warsaw, Poland

ABSTRACT: Without an accurate calculation of the elements of the hydrological balance it is not possible to estimate the magnitude and timing of groundwater recharge which is a major concern in groundwater vulnerability assessment. The hydrological cycle is linked to the processes of weathering, erosion and the biogeochemical cycles which are the main sources of solutes. Due to complex behaviour of the solutes and their migration in soil and rock, a common assumption in vulnerability mapping methodologies is that the pollution migrates in the same manner as water particles. In the study area in the Vistula river valley near Plock, a water balance for the representative soil profile has been calculated using the Thornthwaite (1948) method. Analysis of the balance elements allows determination of periods of water recharge, surplus and deficit in the valley landscape units. From the point of view of solutes transport, the most important are periods of groundwater surplus (months I–III), and recharge (IX–XII). In the first case the, excess water is drained from aquifers to the river network, while in the second conditions are favorable for shallow groundwater contamination by the infiltration process. Groundwater vulnerability mapping by the simplified DRASTIC approach has been verified by geochemical measurements in selected observation wells. The concentrations of Cl and NO_3 have been used as a tracer which may represent the ability of surface pollution to reach groundwater. Atmospheric deposition provides negligible amounts of NO_3. The groundwater contamination level varies depending on the landscape unit. On the Pleistocene terraces under dune fields NO_3 is lowest due to a good recharge ratio. and dilution effect. In all units, the lowest concentrations of NO_3 in groundwater are observed in spring time, due to the dilution effect in the period of water surplus in the fluvial system. These results shows that the obtained vulnerability map represents potential conditions. Map verification can be performed by using geochemical observations, but the location of the monitoring points and time of sampling must be designed to take account of characteristic periods in water balance and of the given landscape units. Groundwater vulnerability maps are the result of modeling, and they should be verified by independent data and observations.

1   INTRODUCTION

Groundwater in river valleys forms important resource used both locally for water supply and regionally contributing through drainage in the base flow of rivers. The environment of the river valley, with floodplain and over-flood terraces, provides some degree of shallow groundwater protection against surface non-point pollution but on the other hand, due to convenient relief and soil conditions, it may be extensively used for agricultural purposes. The groundwater vulnerability to contamination can be described as the derivative of hydrogeological and geomorphological conditions and presented in the form of a carto-graphic model. The result of the modeling is landscape differentiation in to spatial units hav-ing various level of contaminant attenuation potential. The complexity of the vulnerability maps depends on the method applied to their creation. There are three main approaches to making these maps: 1 – expert hydrologic settings, 2 – parametric methods based on matri-ces or rating systems, and 3 – numerical models (Vrba & Zaporozec 1994). The second approach is becoming more popular because it makes it possible to combine hydrogeolog-ical evaluation and automatic rating procedures by using geographical information systems (GIS). The results of modeling do not shows absolute values, but certain relative, dimension-less indices. The resulting groundwater vulnerability maps may present intrinsic (natural) or specific conditions. In both cases the results should be treated as a product of modeling with all its limitations like the effect of scale, simplification and schematization. One important question is to what degree we may believe in modeling results. The answer in the case of mathematical modeling is achieved by model verification, which is usually achieved by the comparison of observed and calculated values.

2   STUDY SITE DESCRIPTION

The lower part of the Vistula river valley near the city of Plock was formed in the late Pleistocene and Holocone. The main form of the valley has been shaped by glacial meltwater outflow, and then by formation of overflood and floodplain terraces. The changing environ-ment in the Holocene, was reflected by the different supply of sediments from the catchment and a run-off regime which caused the transformation of the river channel pattern from braid-ing, to meandering and again to braiding (Mycielska-Dowgiallo & Chormanski 2000). A series of older overflood terraces (Figure 1) from the end of the Pleistocene have been desig-nated as TP, while younger Holocene terraces are designated as TH with the corresponding numbers reflecting their age and sequence (Florek et al. 1987). Characteristic of the TP-1 ter-race are dunes which can be found at the border with the younger terrace TP-2. The Pleistocene terraces have been formed from poorly sorted sediment transported in conditions of braided rivers. The oldest Holocene terrace, TH-1 has been formed since the Sub-Atlantic period, when the subsequent change in hydrological regime of the Vistula River took place. The terrace was formed by a meandering river and then partly buried under sediments of the TH-2 terrace brought by the braided river processes. The TH-3 terrace surface has been dis-sected by flood waters in historical times. The TH-4 terrace has been formed recently between the embankments and the regulation structures. Recent floods are kept within the main channel by dikes which limit the range of inundation. In the Plock area, the flood-plain occupies about 50% of the Vistula river valley floor, which has a width of 4.5 to 9 km. The valley border is a slope of glacial plateau which has a relative elevation 40–50 m above the river. The glacial plateau is built of boulder clay and fluvioglacial sediments.

The hydrology of the river valley is controlled by the flow regime of the main river, by side flow from glacial plateau tributaries, and by the hydrological properties of the floodplain soils and sediments. The average discharge of the Vistula River at the Kepa Polska gauge is 936 m^3/s. The characteristic feature of the Vistula River are winter floods related to the snow melt period (months III–IV) and autumn droughts (months VIII–X). Summer floods origin-ating from rainfall are irregular and occur mostly in months VII–VIII and sometimes in VI or IX. The rainfall floods are shorter than the snow melt floods. Flood waves have

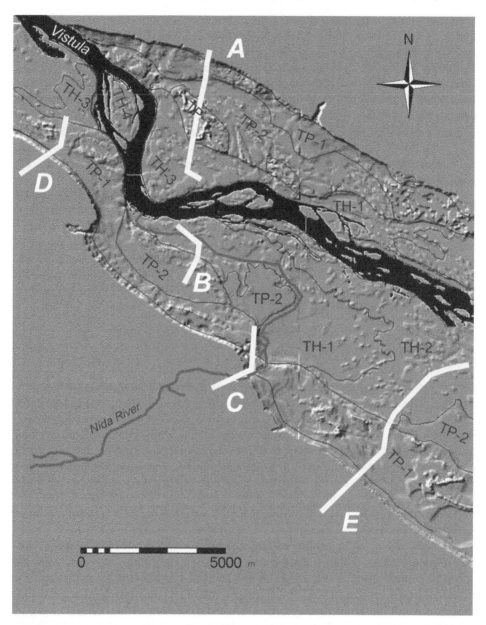

Figure 1. Digital terrain model with terrace descriptions and locations of groundwater monitoring cross-sections.

a peak stage with a relative elevation of 3–5 m (maximum 6 m), and propagation velocity of 3–3.5 km/h.

As a representative of lateral flow from tributaries the small river Nida has been selected. It's minimum measured discharge in the period 1998–2001 was 0.015 m³/s, while the maximum was 0.589 m³/s. For the seasonal distribution of run-off high discharges in the spring time (months III–IV), and low discharges in the late fall (X–XII) are characteristic.

Groundwater in the quaternary sediments of the glacial plateau occurs in two horizons. The deeper aquifer is confined, while shallow groundwater has a free water table. In the valley floor groundwater in the floodplain terraces is shallow and has a free water table at a depth of 2 m. On the over-flood Pleistocene terraces under the dunes the water table is located at a depth of 5–7 m. The studied part of the valley belongs to the system of the main groundwater reservoirs of Poland and has been marked as unit no. 220. The groundwater table in the valley floor reflects the hydraulic conductivity of the floodplain sediments. Groundwater contour lines in the studied area shows that the main direction of groundwater flow is towards the channel of the Vistula River. Seasonal changes of the groundwater table are visible in the whole valley. An increase of groundwater level is observed in months I–IV, with lowest level in months X–XI.

The relationship between the groundwater table in the valley and the Vistula River depends on the hydrological conditions. Using data from groundwater observation sites located on different terraces and from the Kepa Polska river gauge, the correlation coefficient between these variables has been calculated. It is evident that the strength of the relationship depends on dry or wet years, and declines rapidly with distance from the river bank. Strong hydraulic link between river and groundwater occurs only during wet years, and during dry years practically vanish. The maximum range of river influence on groundwater is up to 1,000 m on the flat, permeable Holocene floodplain.

3   GROUNDWATER VULNERABILITY

On the way from the surface to an aquifer, the attenuation of contaminants is controlled by the various geochemical and physical processes occurring in the unsaturated and saturated zones. A common assumption due to the complex behavior of the solutes is that the contaminant migrates in the same manner as the water particles. Only some methods take into account groundwater recharge, which is actually the main driving force in the transport of contaminants from the top of the soil to the groundwater. This important factor is often represented as an average value, calculated as a product of annual precipitation total and coefficient corresponding to surface material permeability. The effective groundwater recharge is controlled by the soil moisture surplus or deficit which is the effect of interaction of the water balance elements.

In the case study of vulnerability map construction a simple method of adding the rankings obtained from two-dimensional tabular classifications performed on digital maps was used (Magnuszewski & Suchozebrski 1998). The selection of essential parameters for classifications is based on DRASTIC approach. DRASTIC was developed originally by the Environment Protection Agency of the USA, and it has been used for evaluating risk of groundwater contamination by nonpoint agricultural sources (Aller et al. 1987). In DRASTIC, parameters are multiplied by weights and added to form a final evaluation of the vulnerability of a given unit.

In this study classification was done in the ILWIS programme using digital thematic maps of the following parameters: depth to groundwater table, recharge conditions (proportional to soil texture), aquifer media (represented by the age of the valley terraces). The groundwater contour map was based on field measurements of 475 farmer's wells in the summer of 1992. Soil texture information comes from digitized 1:25,000 scale soil maps of the area. Aquifer media is a derivative of geomorphological map, showing the age of the terraces.

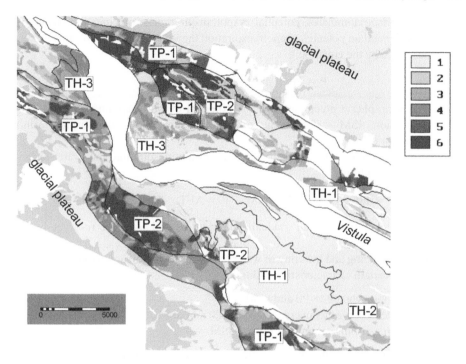

Figure 2.   Results of groundwater vulnerability modeling in the Vistula valley near Plock.

It has been assumed that the most important parameters controlling infiltration processes are depth to groundwater and material permeability in the unsaturated zone. Permeability depends on the texture of the soil and the type of sediments in the bedrock. In the final vulnerability map, a 7-level scale has been applied, starting from 0 – not defined areas to 6 – highly vulnerable areas. Not defined areas are Vistula River islands and also major forest complexes, which are not represented on a soil maps. The map (Figure 2) shows large differences in groundwater vulnerability between glacial plateau, Pleistocene terraces, and Holocene terraces.

4   HYDROLOGICAL CONTROLS

The main direction of groundwater flow in the valley is determined by the drainage of the Vistula River. To this general pattern is added the vertical flux of atmospheric precipitation and infiltration through the unsaturated zone. The vertical flux is responsible for transfer of the contaminants reaching groundwater from non-point sources. The amount and timing of the vertical water flux is controlled by hydrological factors.

The water balance of the soil profile or terrain unit is a controlling factor on the groundwater recharge pattern. To calculate a water balance, we need to know the long term mean monthly meteorological characteristics such as precipitation, evapotranspiration, river run-off, vegetation cover and soil parameters. In the water balance calculation according to the Thornthwaite (1948) method, there are two situation to be considered, depending on which characteristic dominates, potential evapotranspiration or precipitation. When precipitation is higher than potential evapotranspiration the water balance represents wet conditions described by the equation:

$$P = P0 + dSM + R, \quad P0 = AET$$

where: P – atmospheric precipitation, P0 – potential evaportanspiration, AET – aerial evapotranspiration, dSM – changes in soil water storage, R – water surplus forming recharge and run-off.

When precipitation is lower than potential evapotranspiration the water balance represents dry conditions:

$$P = P0 + dSM + D, \quad dSM + P = AET, \quad P0 - AET = D$$

where D – water deficit, which gradually build-up as soil moisture reserves become depleted.

Periods of recovering and lowering of water resources in the soil are marked in the water balance run by certain characteristic conditions:

– period of drying, when P > P0 and dSM < 0
– period of water reserves recovery, when P > P0 and dSM > 0
– period of water surplus, when P > P0 and dSM = 0

In the study area the water balance elements were calculated using the modified (Thornthwaite & Mather 1955) method which takes into account the additional element of river direct run-off. That element has been evaluated for the whole valley area using data from the Nida river catchment, a small tributary of the Vistula river. Relationships between the water balance elements are shown in Figure 3.

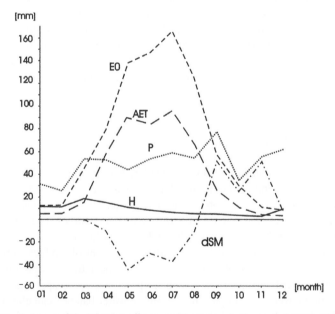

Figure 3.   Water balance of the Vistula valley near Plock area in the period 1990–1994.

From the point of view of solutes transport, the most important are periods of groundwater surplus (months I–III), and recharge (IX–XII). In the first case the excess of water is drained from aquifers to the river network, while in the second conditions are favorable for shallow groundwater contamination by infiltration processes.

5 MODEL VERIFICATION

Verification of groundwater vulnerability maps against actual observations of groundwater quality is difficult because of the long time scale of processes affecting groundwater quality. Long term monitoring programmes of hydrochemical measurements may provide ground truth for the evaluation of vulnerability mapping. In designing such a monitoring network, the groundwater vulnerability map has been used to define the optimal locations of transects of groundwater observation wells. Five transects (A–E) have been created with 24 observation wells, which represent the various geomorphologic units and which cross areas with high gradients of groundwater vulnerability index (Figure 1). To verify groundwater vulnerability in these units, two ions have been selected as a tracers; Cl^- and NO_3^- concentrations. A similar approach has been used by Rupert (2001). Additionally, pH of the groundwater has been measured. Measurements have been made in the years 1996–2000, during which the time of the sampling has been selected according to the characteristic periods marked by the water balance elements.

To evaluate the importance of atmospheric sources of solutes the rainfall chemistry data representing the study area have been investigated. Observations in the years 1994–1996 obtained from the Warsaw University meteorological station at Murzynowo near Plock shows that contamination of rainfall by atmospheric pollution remains low. Extreme values of rainfall pH are in the range of 4.5–6.9. The lowest pH of 4.5–4.8 is characteristic of the late fall and winter period and is caused by emissions from heating installations. For these months, an increase of SO_4^{2+} concentrations has also been observed in the rainfall (up to 18.2 mg/l), and of NO_3^- (up to 1.8 mg/l). The average concentration of NO_3^- in rainfall in the area of the Vistula valley near Plock is 0.68 mg/l (Magnuszewski 2002).

The groundwater hydrochemistry results have been subdivided into sets which represent characteristic periods in the water balance, and also represent given geomorphological units (Figures 4 to 6). Generally the nitrate concentration in the groundwater of the studied area is very high, and in 95% of the observation wells it exceeds the standard for drinking water in Poland of 50 mg/l. The glacial plateau has the worst quality of groundwater, and in some

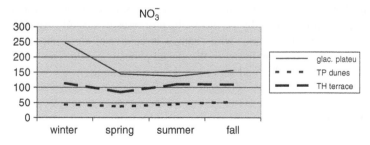

Figure 4.   Concentration of NO_3^- in the groundwater of the Vistula valley in different land units and seasons in mg/l.

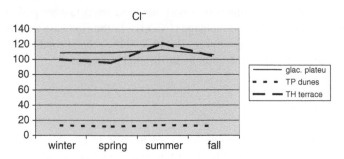

Figure 5.   Concentration of Cl^- in the groundwater of the Vistula valley in different land units and seasons in mg/l.

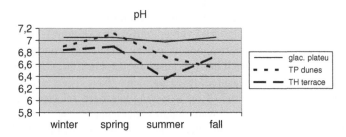

Figure 6.   Values of pH in the groundwater of the Vistula valley in different land units and seasons.

wells an alarming level of 300–450 mg/l has been recorded. These very high concentrations of NO_3 in the groundwater, especially in the glacial plateau, are a result of intensive agriculture and the influence of pollution sources of sanitary origin.

Concentrations of Cl^- and NO_3^- in the groundwater under the permeable soils of the Pleistocene terraces with dunes are the lowest. On the Holocene terraces (places of groundwater discharge) again we observe poorer of groundwater quality. Higher concentrations of Cl^- in the summer period have been observed in 63% of the observation wells. This effect may be due to higher evaportanspiration rates, which affect soil water and also shallow groundwater on the floodplain terraces. The Cl^- concentration under the dunes on the Pleistocene terraces is the lowest, and does not show much variation during the year.

Groundwater pH in all measured wells is in the range 6.0–8.5. In the annual run, in 65% of the wells lowering of pH in summer has been observed. This is probably due to a increase in production of CO_2 by the metabolic processes in the soils. An increase of the pH in the spring time corresponds with changes in acidity of atmospheric rainfall. The most stable pH is that of groundwater in the glacial plateau, and this can be explained by the effect of a carbonate buffer and better insulation of the aquifer.

6   CONCLUSIONS

A study of groundwater flow in the Vistula river valley has shown that drainage towards the main river dominates, together with vertical water exchange dictated by the local water balance. These two components have great influence on groundwater chemistry. The importance of the hydrological cycle for proper selection of sampling periods and the location of the observation points which are used for verification of the groundwater vulnerability maps has been shown. In the water balance, the periods of groundwater surplus and recharge are critical for pollutant migration to and from the aquifer.

The supply of NO_3^- with rainfall is small compared to the other sources, for example agriculture. The degree of groundwater vulnerability can be verified by hydrochemical measurements with special emphasis on Cl^- and NO_3^- concentrations, as well as on pH. The results have shown that all groundwater in the Pleistocene plateau and lower Holocene terraces is highly contaminated by nitrate. This contamination forms an immense pool of nitrogen in the landscape, and will have an influence on the environment and on surface water quality for many years to come.

The groundwater vulnerability map verified by hydrochemical measurements should be interpreted as showing potential rather than actual conditions for contamination. Similar conclusions come from a US EPA study (EPA 1992). Areas of low vulnerability are often contaminated, due to the limited groundwater recharge.

REFERENCES

Aller, L., Bennett, T., Lehr, J., Petty, R. & Hackett, G. 1987. DRASTIC: a standardized system for evaluating ground water pollution potential using hydrogeologic settings. National Water Well Association, Dublin Ohio/EPA Ada, Oklahoma, USA. EPA-600/2-87-035.

EPA. 1992. Another look: National Survey of Pesticides in Drinking Water Wells. Phase II Report. Springfield: Environmental Protection Agency.

Florek, E., Florek, W. & Mycielska-Dowgiallo, E. 1987. Morphogenesis of the Vistula valley between Kepa Polska and Plock in the Late Glacial and Holocene. In: *Evolution of the Vistula river valley during the last 15,000 years*. (Starkel ed.) Part II. Wroclaw: Ossolineum.

Mycielska-Dowgiallo, E. & Chormanski, J. 2000. Evolution of the Vistula valley between Kepa Polska and Plock during the Last Glaciation and Holocene In: *Floodplain pollution control management (Vistula river, Poland)*. (Magnuszewski, Mikulski, F. Brinkmann eds.), Koblenz: IHP/OHP.

Magnuszewski, A. 2002. Systemy geoinforamcyjne w badaniach ekohydrologicznych. Przykład doliny Wisły pod Płockiem. WGSR Uniwersytet Warszawski, Warszawa.

Magnuszewski, A. & Suchozebrski, J. 1998. Wykorzystanie modelu DRASTIC do okreslenia stopnia potencjalnego zanieczyszczenia wod podziemnych w dolinie Wisly pod Plockiem. In: *Hydrologia u progu XXI wieku*. (Magnuszewski, Soczynska eds.), Polskie Towarzystwo Geograficzne. Warszawa: Retro Art.

Rupert, M.G. 2001. Calibration of the DRASTIC ground water vulnerability mapping. method. *Ground Water* 39(4): 625–630.

Thornthwaite, C.W. 1948. An approach toward a rational classification of climate. *Geographical Review* 38(1): 55–94.

Thornthwaite, C.W. & Mather, J.R. 1955. The water balance. Publications in Climatology. Vol. VIII, no. 1. Center for Climatic Research, University of Delaware.

Vrba, J. & Zaporozec, A. 1994. Guidebook on mapping ground water vulnerability. International Association of Hydrogeologists. Vol. 16. Hannover: Verlag Heinz.

CHAPTER 14

Modeling and mapping groundwater protection priorities using GIS: the case of Dar Es Salaam city, Tanzania

R.R.A.M. Mato

Department of Environmental Engineering, University College of Lands and
Architectural Studies (UCLAS), Dar es Salaam, Tanzania

ABSTRACT: The Dar es Salaam city has a population of about 2.5 million, growing at about 8% per year. Water supplies come from three surface water treatment plants built on rivers. The system is under-capacitated, supplying only about 50% of the demand, a situation that makes groundwater an alternative for augmenting the supply. The water authority has drilled about 40 boreholes. In addition, there are increasing numbers of private and community's boreholes and shallow and dug wells. However, the potential of groundwater as a long-term source of water supplies is being undermined by urban activities, like crude practices of waste disposal and sprawling of unplanned settlements, which threaten its quality. Such practices expose population to serious health risks especially the urban poor. Sustainable utilization of groundwater resources in the city is thus inevitable and pro-active means are considered appropriate. An empirical model has been developed and used as a tool to delineate groundwater protection priorities in the city. It consists of five factors: water quality, yield, vulnerability, use value of groundwater and land-use. It uses a system of weighting and rating of the factors and aggregates them in a single number, the protection index that shows relative protection priority. The model was used in a GIS environment that enabled the mapping of groundwater protection priorities for Dar es Salaam. The results showed that Charambe and Mbagala wards have high priority for protecting groundwater, while Manzese and Tandale have low. The prioritization enables rational resource mobilization and long-term groundwater management planning. Planners in allocation of land for various uses can easily adopt the model. Moreover, the model has become a first practical step towards sustainable management of groundwater in the city of Dar es Salaam.

1 INTRODUCTION

Groundwater quality protection has increasingly become a public issue worldwide, and it aims at preserving the quality and the quantity of groundwater in agreement with other interests of land use (Granlund et al. 1994). In many countries, a groundwater protection plan is used as a guide in supervision of water legislation, issuing of permits, planning land use etc. The need for protecting aquifers arises after detection of increased demands for groundwater resources development and usage, and the recognition of deteriorating quality resulting from pollution sources. Cities and many other urban centres have increasingly become major sources of diffuse pollution of groundwater. Reported cases of aquifer degradation in urban environments are on the increase in literature (Morris et al. 1994; Chilton 1999; Sililo ed. 2000; Foster 1999; 2001; Pokrajac 2001). The aquifer degradation results from inadequate control of groundwater exploitation and indiscriminate disposal of wastes on or beneath land and other urbanization related human activities. This degradation has contributed to escalating water supply costs, increasing water resource scarcity and growing health hazards (Morris et al. 1997). In low-income countries (Tanzania inclusive), where both financial and technical resources are scarce while experiencing urban population explosions, there exists a challenge of protecting groundwater as a cheap and more readily available resource. The protection protocol is based on the principle that prevention is less expensive than remediation of polluted aquifer, which is a costly, long term and technically demanding task if not impossible.

The city of Dar es Salaam is the largest and most populous urban centre in Tanzania. Its estimated population is 2.5 million (about 30% of the urban population in Tanzania), growing at a rate of about 8% per year (Baruti et al. 1992). There is acute deficiency in infrastructure provision: housing, water supply, sanitation, transportation and energy. Water supplies come from three surface water treatment plants (Upper Ruvu, Lower Ruvu and Mtoni) constructed more than 25 years ago. The plants are operating at about 60% of the installed capacity (JICA, 1991), supplying only about 48% of the city water demand (Mato 2002). The remaining 52% of the demand is augmented by groundwater source, normally exploited through shallow and deep wells (Mato 2000). More than 800 boreholes with depth up to more than 100 m and yield of up to 60 m^3/h have been drilled in the city (DDCA 2000). There are numerous shallow wells in areas without piped water (e.g. Kiwalani, Mbagala, Ukonga etc) and where supply is rationed (Mato et al. 2000; Kjellen 2000). In this respect, groundwater has increasingly become an important and a principal water source for the city residents. However, this resource is not well managed as the exploitation is uncontrolled and wastes are indiscriminately disposed of on or underneath the land. Such practices are threatening the usefulness of groundwater resource for the city of Dar es Salaam, and if left unattended can lead to increased water costs, epidemics (such as cholera, typhoid, cancer and other unknown diseases) and cause adverse impacts to other terrestrial and aquatic ecosystems.

Urban residential districts without or with incomplete, coverage by sewerage, seepage from on-site sanitation systems such as pit latrines and septic tanks, presents the most widespread and serious diffuse pollution sources in Dar es Salaam. About 90% of the population use pit latrines and septic tanks for sanitation (Chaggu et al. 1994). Most of the urban inhabitants (60–70%) live in unplanned (squatter) areas, where there are little or no infrastructure services for waste collection (National Environment Management Council 1995; Ministry of Lands and Human Settlement 2000). Due to increased urban population, the quantities of solid wastes to be collected and transported for disposal have also increased tremendously in recent years (Mato & Kaseva 1999; Kaseva & Mbuligwe 2004). Only about 20–40% of the urban population receive regular solid waste collection services (especially after privatisation) in most cases confined to few areas, usually the urban centres and high-income neighbourhoods (Mato 1997; Lussuga & Yhdego 1997; Kaseva & Mbuligwe 2004). The uncollected wastes (60–80%) are mainly buried at generation sites, many of them less than 1.0 m below ground level, hence becoming a diffuse source of pollution. The collected wastes are disposed of crudely at "dumpsites" (abandoned Tabata and Vingunguti, and now Mtoni), where there are no safe means of handling leachate (heterogeneous concentrated contaminant containing toxic substances).

Other source of groundwater pollution in Dar es Salaam can emanate from industries, which have been operating with inadequate or without waste treatment facilities, some for more than 40 years. Mato (2002) estimated pollution from Dar es Salaam based industries in 2,715 tons/day and 15,454 tons/day of biochemical oxygen demand (BOD_5) and suspended solids (SS) respectively. It is estimated that at least 15–20% of this loading may be reaching the aquifers.

Urban agriculture (one of the fast growing informal activity) and petroleum products (from filling stations, depots and garages) also constitute emerging sources of groundwater pollution in the city. Organic manure, synthetic fertilizers, agrochemicals used in urban agriculture (including DDT) and petroleum hydrocarbons penetrates into the aquifers. Though at the moment there is no monitoring system in place there are risks associated with these substances.

With the above background, a research to assess the groundwater pollution in Dar es Salaam City was conducted by the University of Dar es Salaam, Tanzania in collaboration with the Technical University Eindhoven, The Netherlands from 1997–2002. Among other things, our research involved identification of major sources of pollution and developing protection strategy of groundwater in the city. Model to assist in determining protection priorities was developed and is the basis of this paper.

2   MATERIALS AND METHODS

The methodologies used in this research were mainly laboratory analysis and modeling. At first, a survey of major potential pollution sources was conducted and a quantification of pollution loads made. A database (spatial and non-spatial) of boreholes and petroleum filling stations was established. The geographical references of boreholes and petroleum filling stations were picked with Global Positioning System (GPS) handsets (Garmin II make).

During the research period, monitoring of groundwater quality was conducted. Both inorganic and organic characteristics were investigated, including nitrate, total organic carbon, chemical oxygen demand (COD), total dissolved solids (TDS), chloride, sulphates, total oil content and polyaromatic hydrocarbons (PAH). Standard methods were used to analyse inorganic parameters.

A groundwater vulnerability assessment was then conducted using the DRASTIC model. Finally, a rapid assessment model to assist in the groundwater protection strategy was developed. The Geographical Information System (GIS) software (ArcView 3.1) was used for mapping.

3   DEVELOPMENT OF GROUNDWATER PROTECTION STRATEGY

The principal goal of a groundwater protection strategy is to conserve the groundwater resource by preventing/reducing quality deterioration and overexploitation. Recognising that the groundwater quality in Dar es Salaam is deteriorating (Mato 2002) and that there are many potential pollution sources (as discussed above) a protection plan was found inevitable. Literature have many general recommendations for achieving a good groundwater management scheme; including pollution sources elimination/minimization, setting monitoring networks, interagency coordination etc (Cramer & Vrba 1987; Foster 1985; 1987; Pokrajac 1999; Soetrisno 1999; Vujasinovic et al. 1999). For the city of Dar es Salaam, a protection plan developed included development of groundwater vulnerability map and a model to assess protection priorities.

3.1   *Groundwater Vulnerability Map for Dar es Salaam City*

Assessment for groundwater vulnerability has now become a common tool in groundwater management and protection, as is the case of monitoring, modeling and mapping (Canter 1985; Cramer & Vrba 1987; USA National Research Council 1993; Palmer et al. 1995; Pebesma & Kwaadsteniet 1997; Kelly & Lunn 1999; Ibe et al. 2001). The vulnerability assessment and vulnerability maps represent an important preliminary tool in decision-making pertaining to the management of groundwater quality. They provide a useful framework within which to designate priorities for the implementation of pollution protection and control measures, though field investigations and monitoring are not ruled out (Blau 1981; Gowler 1983; Foster 1985; Aller et al. 1985, Anderson & Gosk 1987). The vulnerability maps also serve to inform and educate the public, because non-professional people can readily understand their concept. They also create public awareness about potential pollution problems of groundwater, a situation needed for effective implementation of future protection programmes.

The groundwater vulnerability map for the city of Dar es Salaam was developed using the DRASTIC model coupled to GIS software. The DRASTIC model (Eq. 1) is an empirical groundwater model that estimates groundwater contamination vulnerability of aquifer systems based on the hydrogeological settings of that area (Aller et al. 1985; Aller et al. 1987). It employs a numerical ranking system that assigns relative weights to various parameters. The acronym DRASTIC is derived from the seven factors considered in the method, which are **D**epth to groundwater [**D**], net **r**echarge rate [**R**], **A**quifer media [**A**], **S**oil media [**S**], **T**opography [**T**], **I**mpact of the vadoze zone [**I**], and **C**onductivity (hydraulic) of aquifer [**C**]. The final results for each hydrogeological setting are a numerical value, called DRASTIC index. The higher the value is, the more susceptible the area in question is to groundwater pollution. The DRASTIC input data were obtained from more than 472 borehole drilling data availed by the Dar es Salaam Drilling and Construction Company (DDCA). Other DRASTIC parameters were obtained from the topographical map of Dar es Salaam.

$$DI = D_w D_r + R_w R_r + A_w A_r + S_w S_r + T_w T_r + I_w I_r + C_w C_r \tag{1}$$

The vulnerability map (Figure 1), divided the city into low, moderate and high vulnerability zones. The map shows that about 50% of the city falls under the high vulnerability zone. Industrial and high-density residential areas are all located in the "red zones", that is, high and very high vulnerability classes. About 30 samples were taken from boreholes located in different vulnerability classes to verify the results of the DRASTIC model simulations. The nitrate concentrations in different vulnerability classes showed good correlation to the DRASTIC model predictions. Considering that over 90% of the Dar es Salaam inhabitants are using on-site sanitation facilities, nitrate levels in groundwater can be used as an indicator of aquifer quality status in the city, upon which pollution potential can be based. The groundwater vulnerability map developed for the City of Dar es Salaam marks a milestone in efforts to combat pollution of drinking water sources. The map can be used as a general guidance to groundwater pollution control strategies.

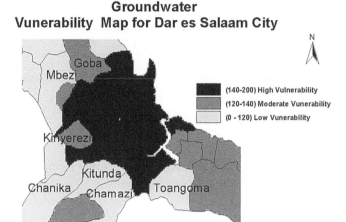

Figure 1.   Groundwater vulnerability map for Dar es Salaam City, Tanzania (2002).

Table 1.   Weights of factors in the WYVUL model.

Factor	Weight
Water quality	5
Yield of aquifers	4
Vulnerability of groundwater to pollution	3
Use value of groundwater	2
Land-use characteristics	1

3.2   *Groundwater Protection Assessment Model*

3.2.1   *Model concept*

Form this study, a rapid assessment model, the *WYVUL model* (Eq. 2), for assessing groundwater protection priorities was formulated and applied. The philosophy is to find ways in which groundwater protection and other urban activities can co-exist.

The model, *WYVUL*, is composed of five factors, which are:

- Water quality, W
- Yield of the aquifers, Y
- Vulnerability index of groundwater to pollution, V
- Use value of the groundwater, U
- Land-use characteristics, L

The WYVUL factors were given weights in accordance with the recognized importance in dictating groundwater protection of a certain location within Dar es Salaam area. The weights have a range of 1–5 and are shown in Table 1. The weights were assigned basing on the technical knowledge of the potential of the factors to influence groundwater management plans. Different rating scheme was applied to every factor, with scores of 1–10.

The overall concept behind the WYVUL model is *"to protect clean aquifers with huge amounts of water located in vulnerable areas, where groundwater is of high demand*

Table 2.   Rating for water quality factor in the
WYVUL model.

Nitrate concentration range (mg/l)	Rating
<20	10
21–50	8
51–100	6
101–120	4
Above 120	1

particularly places with very low physical development". Areas fulfilling this are having high protection index and hence greater protection priority.

$$PI = W_wW_r + Y_wY_r + V_wV_r + U_wU_r + L_wL_r \qquad (2)$$

where *PI* is the protection index and the subscripts, *w* and *r* denote assigned weight and rating of each factor. The lower and upper limits for *PI* are 5 and 500 respectively.

3.3   *Description of WYVUL model parameters*

Water quality: The water quality defines the types and amounts of substances contained (in suspended or dissolved form) in groundwater. The water quality factor is regarded as a determining factor in the use of groundwater. The principle is to protect groundwater that has not yet been polluted. The water quality factor can be derived represented by one pollution parameter or an aggregate of many parameters. For this study, the nitrate concentration was taken as a principal component in assessing the water quality of the groundwater. The reason for choosing nitrate is because of rampant use of on-site sanitation systems in the city, which can lead to nitrate contamination of the groundwater. The water quality factor was given a weight of 5 in the model (this is the highest weight value). A rating system for water quality was established as shown in Table 2. The WHO and Tanzania nitrate standards of 50 mg/l and 100 mg/l respectively, were taken as reference values for establishing the ratings. The highest water quality score is 10 and was awarded to an area where the nitrate concentration in the groundwater was below 20 mg/l, a value considered to represent clean aquifer (Freeze & Cherry 1979). Ratings of 8 and 6 were applied where the nitrate levels were 21–50 mg/l and 51–100 mg/l respectively. The minimum rating of 1 was applied to an area where nitrate concentration was above 120 mg/l, a value far above the Tanzanian standard, for which the state of the aquifer was considered to be very poor or greatly damaged.

   Yield of aquifers: Aquifer yield is defined as the maximum rate of withdrawal of water that can be sustained by an aquifer without causing an unacceptable decline in the hydraulic head in the aquifer (Freeze & Cherry 1979). The yield factor in the WYVUL model expresses the available amounts of groundwater that needs protection. It was considered the second important factor in the groundwater protection scenario, with a weight of 4 (Table 1). The *WYVUL* model considers it being most logical to protect unpolluted aquifers with high yields. The philosophy is that the larger the yield of aquifers the higher the protection priority. Accordingly, ratings for aquifer yield were formulated as shown in Table 3. The highest score is 10 and is awarded to aquifers with yield of more than 40 m³/h, sufficient to develop a public water supply scheme.

Table 3.   Rating for the yield of aquifer factor in the *WYVUL* model.

Yield Range (m³/h)	Rating
<5	2
5–10	4
11–20	6
21–40	8
Above 40	10

Table 4.   Rating for the vulnerability factor in the *WYVUL* model.

Vulnerability (DRASTIC Index)	Rating
Low (<120)	2–4
Moderate (121–140)	5–7
High (>140)	8–10

Table 5.   Rating for use value of groundwater factor in the *WYVUL* model.

Water supply status	Rating
Groundwater (GW) not used as alternative source	1
With piped water supply (service over 4 days per week), GW alternative source	6
With piped water supply (service less than 4 days per week), GW used as alternative source	8
Without piped supply, GW used as alternative source	10

Vulnerability of groundwater to pollution: This factor takes care of the intrinsic hydro-geological characteristics of the area and its potential to pollution. Here, priority for protection is given to an area which has good groundwater quality, high yield from the aquifers and vulnerable to pollution. The vulnerability of groundwater to pollution factor was considered third important and was given a weight of 3 (Table 1). The ratings for this factor were formulated from the groundwater vulnerability map of Dar es Salaam City (Figure 1). The rating has a range of 1–10, with the highest score of 8–10 when the area to be protected falls within the high vulnerable zone (Table 4). The minimum score of 2–4 was given to a low vulnerable zone, for which it was considered that the existing natural protection due to soil profile could sufficiently attenuate the pollution loads before reaching the groundwater.

Use value of groundwater: This factor expresses the value of groundwater to communities in specific location in Dar es Salaam. The general proposition is that areas without piped water supplies depend directly on groundwater as alternative source. For instance areas of Kiwalani, Kipunguni depend largely on groundwater. In such areas where water supply from piped water is very low or in-existent, the groundwater is considered to have a higher use value and therefore higher protection priority. In the WYVUL model, the use value of groundwater is given a weight of 2 (Table 1), and ratings range from 1–10 as shown in Table 5. Alternatively, it was considered that since

Table 6.   Rating for landuse characteristics factor in the *WYVUL* model

Landuse Characteristics	Rating
Rural-urban areas (urban fringe areas)	10
Commercial or institutional	5
Planned residential areas	2
Unplanned residential and industrial areas	1

Table 7.   Protection priority categorization.

Protection index range	Protection priority category
1–80	Low
80–100	Moderate
100–200	High

groundwater is used to augment public water supplies for the entire Dar es Salaam City, it has a constant use value. Both alternatives were considered in computing the protection indices.

Land Use Characteristics: In the WYVUL model, the type of human activities or urban infrastructure on a certain geographical areas is referred to as land use characteristics. This factor takes care of the level of physical development of the areas, which can give difficulties in varying degrees to protection programmes, and it is given a weight of 1. The model considers that the less the physical development which is in place, the easier may be to execute protection programmes. A rating with a range of 1–10 has been formulated with urban-rural areas (rural areas within the broader Dar es Salaam City environment e.g. Charambe and parts of Mbagala wards etc) having the highest score of 10 (Table 6). The argument is that it is easier to take groundwater protection measures in the urban-rural settings than in a built-up urban area e.g. unplanned areas of Manzese, Tandale, Keko etc. Therefore, unplanned "squatter" areas were assigned a rating of 1.

3.3.1   Protection index computation and GIS coupling

The protection index was computed for each borehole location (where all the parameters were known). The borehole location was taken as a node to represent the surrounding area. The protection index was between 50 and 200, of which over 50% were above 100. The higher the relative number of protection index the higher the priority the area has for being protected as a potential groundwater source. A categorization scheme was adopted (Table 7) in which three zones: low, moderate and high protection priorities were identified. For further visualization of the protection index concept, mapping of the indices was carried out using the GIS software (ArcView -version 3.1). The boreholes location provided the geographical references needed for GIS analysis, from which the groundwater protection priority map was produced (Figure 2).

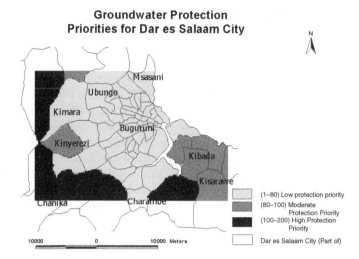

Figure 2.   Groundwater protection priority map for Dar es Salaam City (2002).

4   RESULTS AND DISCUSSION

4.1   *Groundwater protection priority map*

The groundwater protection priority map (Figure 2) delineates areas of Dar es Salaam with varying degree of groundwater protection importance. The map shows that major parts of Charambe, Mbagala, Ukonga and Yombo Vituka wards have high protection priorities. The zone of high protection priorities also includes the built-up areas of City Centre and parts of Upanga and Kariakoo. These areas may have high status because of the presence of a sewerage system, that drains away most of the nitrate sources from reaching the ground-water. The remaining parts of the city fall under moderate priority zone except for few areas, like Magomeni, Manzese, Kinondoni, Mwananyamala and Mtoni, which are delin-eated as low priority.

A sensitivity analysis was performed by changing the weights of key parameters in the model like water quality, yield, vulnerability and landuse characteristics. The results of the sensitivity analysis indicate that water quality, W; yield, Y; vulnerability, V and landuse characteristics, L are all important factors in the model. The analysis indicated that water quality and landuse are on two extreme ends, each pulling toward low and high protection priority respectively. These two factors do dictate in determining the protection priority. In more general terms it means that the status of the groundwater quality and the level of physical development of the area may dictate the level of protection strategy to be applied. However, Mbagala and Charambe ward areas were outstandingly rated high protection zones by all the sensitivity analysis scenarios.

4.2   *Model validation*

Since the protection index is a relative and immeasurable quantity, it cannot be directly measured in the field. Validation of the model to some extent was made through the knowledge of the areas. The areas of Mbagala and parts of Charambe are places with high aquifer yield (the DDCA regard these areas as "well field"), relatively good quality water, low physical development (mostly rural-urban fringe), with sand soils and very low water provision. Such areas having a high priority of protection is in agreement with the WYVUL concept, "to protect clean aquifers with huge amounts of water located in vulnerable areas, where groundwater is of high demand particularly places with very low physical development". The Recharge areas of Ukonga ward (Service Plan, 1997) are in the high protection priority zone. In these areas physical development is still low, and sand soils prevail. Upanga area and City Centre, which are rated high priority, show considerably high yield, low nitrate concentration (may be because most areas are sewered) and sand soils; however, physical development is relatively high. Again, this situation indicates the high pulling power of the water quality factor in the model. However, such areas (City Centre) can be included in groundwater protection programmes at regional scale. Comparatively, unplanned "squatter" residential districts with rampant use of on-site sanitation facilities and shallow groundwater dominate areas like Manzese and Mtoni designated as low protection priority zones.

4.3   *Model use and application*

The *WYVUL* model results give indication of areas that need protection so as to achieve sustainable groundwater utilization in Dar es Salaam city. The results can be used by physical planners (during land allocations) and groundwater resource managers, the Central Water Board, the custodian of water resources according to the Water Utilization (Control and Regulation) Act. The maps can also be used by drillers (like DDCA) to guide their drilling plans and communities to understand the urgency needed to safeguard the water they depend on. However, more data inputs (data sets from the field) in the model will refine the protection priorities demarcation. Noting that the environment is changing over time, the protection priorities maps also need to be updated with time to accommodate the changes that take place. An aggregate of parameters for the water quality, W, factor in model may be a good representation of the actual status of groundwater quality in Dar es Salaam. To many programmes, the effort has been directed towards managing the infrastructure, leaving out the resource base. The prioritization enables rational resource mobilization and long-term groundwater management planning. Moreover, the model has become a first practical step towards sustainable management of groundwater in the city of Dar es Salaam.

5   CONCLUSION

Potential groundwater pollution sources exist in Dar es Salaam City, which threaten to severely degrade the aquifers. A groundwater management strategy is thus an essential element for achieving sustainable utilization of the groundwater resource. The WYVUL model can be used to delineate areas with varying protection priorities. The model identified Mbagala, Ukonga, Yombo Vituka and Charambe areas as having high priority for protection at the present time. With appropriate modifications (to suit local conditions, the model can be used in many parts in the world. For the city of Dar es Salaam, the model has become a first practical step towards sustainable groundwater management.

ACKNOWLEDGEMENTS

I wish to thank Prof. Dr. F. J .J. G. Janssen of the Environmental Technology Group, Faculty of Chemistry and Chemical engineering at the Technical University Eindhoven, the Netherlands for his tireless supervision during my research. The same compliments goes to Prof. J.H.Y. Katima of the Department of chemical Engineering, University of Dar es Salaam. I also extend my appreciation to the governments of Tanzania and The Netherlands for financing the research.

REFERENCES

Aller, L., Bennett, T., Lehr, J.H. & Petty, R.J. 1985. DRASTIC: a standardized system for evaluating groundwater pollution potential using hydrogeological settings. US-EPA 600/2-85-018. Ada, Oklahoma USA.

Aller, L., Bennett, T., Lehr, J.H., Petty, R.J. & Hackett, G. 1987. DRASTIC: a standardized system for evaluating ground water pollution potential using hydrogeological settings. US-EPA Reports, 600/2-87-035. Washington DC, USA.

Anderson, L.J. & Gosk, E. 1987. Application of vulnerability. In: *"Vulnerability of soils and groundwater pollutants" Proceedings and Information* No.38. (van Duijvenbooden & van Waegeningh eds.). TNO Committee on Hydrological Research, The Hague, Netherlands: 309–320.

Baruti,P., Kyessi, A.G., Rugumamu, W., Majani, B., Halla, F., Kironde, J.M.L. & Chaggu, E.J. 1992. Managing the sustainable growth and development of Dar es Salaam: Environmental profile of the metropolitan area. United Nations Center for Human Settlements (HABITAT), URT/90/033.

Blau, R.V. 1981. Protection areas, a special case of groundwater protection. *Science of Total Environment*, 21: 363–372.

Canter, L.W. 1985. Methods for Assessment of Groundwater Pollution Potential, In: *Groundwater Quality* (Ward, Giger & McCarty eds.) , John Wiley and Sons, Inc., New York: 270–306.

Chaggu, E.J., Mgana, S., Mato, R.R.A.M., Kassenga, G.K. & Rwegasira, M. 1994. The investigation of groundwater pollution in Majumbasita area. Dar es Salaam, Tanzania. Research sponsored by IDRC.

Chilton, J. (ed.) 1999. Groundwater in Urban Environment: Selected City Profiles. Rotterdam: Balkema.

Cramer, W. & Vrba, J., 1987. Vulnerability mapping. In: *"Vulnerability of soils and groundwater pollutant" Proceedings and Information* No.38 (van Duijvenbooden & van Waegeningh eds.). TNO Committee on Hydrological Research, The Hague, Netherlands: 45–47.

Drilling & Dam Construction Agency (DDCA). 2001. Working files. Dar es Salaam: Maji-Ubungo.

Foster, S.S.D. 1985. Groundwater pollution protection in developing countries. *Hydrogeology*, 6: 167–200.

Foster, S.S.D. 1987. Fundamental concepts in aquifer vulnerability, pollution risk and protection strategy. In: *"Vulnerability of soils and groundwater pollutants" Proceedings and Information* No.38 (van Duijvenbooden & van Waegeningh eds.). TNO Committee on Hydrological Research, The Hague, Netherlands: 45–47.

Foster, S.S.D., 2001. The interdependence of groundwater and urbanization in rapidly developing cities. *UrbanWater*, 3: 209–216.

Foster, S.S.D., Morris, B.L., Lawrence, A.R. & Chilton, J. 1999. Groundwater impacts and issues in developing cities – An introductory review. In: *Groundwater in Urban Environment* (Chilton ed.), Selected City Profiles. Rotterdam, Balkema: 3–16.

Freeze, E.A. & Cherry, J.A. 1979. Groundwater, Pretence-Hall Inc., Englewood Cliffs, N.J.

Gowler, A. 1983. Underground purification capacity. *IAHS publications*, 142(2): 1063–1072.

Granlund, K.A., Nysten, T.H. & Rintala, J.P. 1994. Protection plan for an important groundwater area – a model approach. *IAHS Publ.* No. 220: 393–398.

Ibe, K.M., Nwankwor, G.I. & Onyekuru, S.O. 2001. Assessment of groundwater vulnerability and its application to the development of protection strategy for water supply aquifer in Owerri, Southeastern Nigeria. *Environmental monitoring and assessment*, 67: 323–360.

Japan International Cooperation Agency (JICA). 1991. Report on rehabilitation of Dar es Salaam water supply in the United Republic of Tanzania.

Kaseva, M.E. & Mbuligwe, S.E. 2004. Appraisal of solid waste collection following private sector involvement in Dar es Salaam City, Tanzania. Habitat International. In press.

Kelly, C. & Lunn, R. 1999. Development of a contaminated land assessment system based on hazard to surface water bodies. *Water Research*, 33(6): 1377–1386.

Kjellen, M. 2000. Complementary water systems in Dar es Salaam, Tanzania: The case of water vending. *International journal of water resources development*, 16(1): 143–154.

Kongola, L.R.E., Nsanya, G. & Sadiki, H. 1999. Groundwater resources: development and management, an input to the Water Resources Management Policy Review (Draft), Dar es Salaam, Tanzania.

Lussuga Kironde, J.M. & Yhdego, M. 1995. The governance of waste management in urban Tanzania: towards a community based approach. *Resources, Conservation and Recycling*, 21(4): 213–226.

Mato, R.R.A.M. 2000. Solid waste management practices: The case of Tanzania urban centres. Proc. of the workshop on recycling and safe solid waste disposal for municipal areas in Tanzania: Envitech, Dar es Salaam, Tanzania.

Mato, R.R.A.M. 2002. Groundwater Pollution in Urban Dar es Salaam, Tanzania: Assessing Vulnerability and Protection Priorities. Eindhoven University of Technology, The Netherlands.

Mato, R.R.A.M. & Kaseva, M.E. 1999. Critical review of Industrial and Medical Solid Wastes Handling Practices in Dar es Salaam City. *Resources Conservation and Recycling*, 25: 271–287.

Ministry of Lands and Human Settlements Development. 2000. National Human Settlements Development Policy. Dar es Salaam, The United Republic of Tanzania.

Morris B.L., Lawrence, A.R. & Stuart, M.E. 1994. The impact of urbanization on groundwater quality. British Geological Survey, Technical Report WC/$/56. Keyworth, UK.

Morris, B.L., Lawrence, A.R. & Foster, S.S.D. 1997. Sustainable groundwater management for fast growing cities: Mission achievable or mission impossible?. In: *Proceedings of the XXVII IAH Congress "Groundwater in the urban Environment"* (Chiton et al. eds.), Vol.1 Problems, Processes and Management, Nottingham, UK, A A Balkema, Rotterdam: 55–66.

National Environment Management Council (NEMC). 1995. Tanzania National Conservation Strategy for Sustainable Development.

Palmer, R.C., Holman, I.P., Lewis, M.A. & Robins, N.S. 1995. Guide to Groundwater Vulnerability Mapping in England and Wales: Soil survey and Land Research Centre, London.

Pebesma, E.J. & Kwaadsteniet, J.W. 1997. Mapping groundwater quality in the Netherlands. *Journal of Hydrology*, 200: 364–386.

Pokrajac, D. 2001. Groundwater in urban areas. *Urban Water* 3(3): 155.

Pokrajac, D. 1999. Interrelation of wastewater and groundwater management in the city of Bijeljina in Bosnia. *Urban Water* 1(3): 243–255.

Sililo, O. (ed.) 2000. Groundwater: Past Achievements and Future Challenges. Proceedings of the IHA Conference, Cape Town, South Africa. Rotterdam: Balkema.

Service Plan 1997. Evaluation of groundwater sources of Dar es Salaam City. Consultancy Report to World Bank.

Soetrisno, S. 1999. Groundwater management problems: Comparative city case studies of Jakarta and Bandung, Indonesia. In: *Groundwater in the Urban Environment* (Chilton ed.), Vol. 2. Rotterdam, Balkema: 63–68.

USA National Research Council. 1993. Groundwater Vulnerability Assessment: National Academy Press, Washington D.C.

Vujasinovic, S., Matic, I., Lozajic, A. & Dasic, M. 1999. Hydrogeological problems concerning the multipurpose use of Belgrade's groundwater resources. In: *Groundwater in the Urban Environment* (Chilton ed.), Vol. 2Rotterdam: Balkema: 263–266.

CHAPTER 15

Vulnerability mapping in two coastal detrital aquifers in South Spain and North Morocco

J. M. Vías [1], M. Draoui [2], B. Andreo [3], A. Maate [2], J. Stitou [2],
F. Carrasco [3], K. Targuisti,[2] & M.J. Perles [1]

[1]*Departamento de Geografía, Facultad de Filosofía y Letras, Universidad de Málaga, Málaga, Spain*
[2]*Université Abdelmalek Essaâdi, Département de Géologie, Faculté des Sciences, Tétouan, Morocco*
[3]*Departamento de Geologia, Facultad de Ciencias, Universidad de Málaga, Malaga, Spain*

ABSTRACT: This paper describes how the GOD method is applied to vulnerability mapping of multilayer aquifers beneath the Martil-Alila river (North Morocco) and the Velez river (South Spain). These aquifers are exposed to nitrate and faecal-coliform contamination due to human activity. In both aquifers, the G factor (groundwater occurrence) was found to be influential in assessing the degree and distribution of vulnerability. In some sectors of the aquifers, outcropping marls, silts or clays imply decreased vulnerability, in comparison with the unconfined sector of the aquifer. Only the materials that outcrop in surface of the aquifer were evaluated because vulnerability for the whole multilayer aquifer cannot be assessed by the GOD method. This method can only assess vulnerability for a single layer because it only takes one value of the depth-to-groundwater-table factor into account.

1 INTRODUCTION

The considerable influence of urban growth on groundwater resources and their exploitation has severely affected coastal aquifers. This increased urban pressure, moreover, implies the potential danger of groundwater contamination from, for example, solid and liquid urban waste, pesticides and fertilisers, as well as the risk of over-exploitation and the subsequent salinization of the aquifer.

In response to this situation, a project has been established to study aquifer vulnerability, as a tool for future protection strategies. The goal of the project is to determine the degree of protection required by aquifers, by applying various methods to evaluate their vulnerability.

As part of the above mentioned project, this paper presents the initial results obtained from the vulnerability mapping of the Martil-Alila and Vélez detritic aquifers. These present similar characteristics in such aspects as hydrogeological functioning, lithology and environment-related contamination risks. Thus, both aquifers have common features on the presence of some contaminants coming from urban and agricultural activities including seawater intrusion. The future development of the main metropolitan areas (Tetouan and Vélez Málaga), particularly in their respective coastal urban settlement, for touristy use, could increase the impact on the groundwater quality.

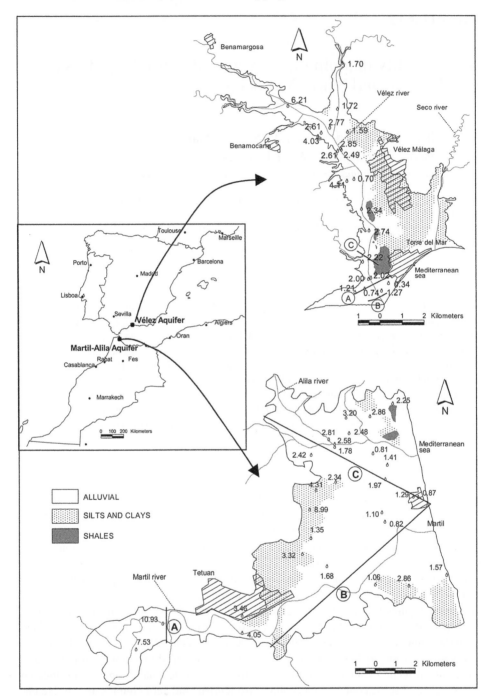

Figure 1.   Location of the Vélez and Martil-Alila aquifers. Hydrogeological map (thickness values of the unsaturated zone in m) and location of geological cross-sections (A, B and C).

The Vélez aquifer is in South Spain (Figure 1), 30 km east of Málaga, and has a total surface area of 20 km². The Martil-Alila aquifer is in North Morocco (Figure 1), near Tetouan, and has a total surface area of 90 km².

The Vélez aquifer is constituted of Quaternary alluvial detritus, mainly gravel, sand, silt and clay and Pliocene marls and conglomerates. The substrate and borders of the aquifer are comprised of metapelitic Palaeozoic materials. In the delta sector, the alluvial sequence shows a level of silt 5 m thick that has led to the existence of two aquifers (Figure 2), one superficial and unconfined (15 m thick), and the other, deep and semi-confined (García Arostegui 1998).

Mean annual precipitation in the area is 630 mm, although with the marked interannual irregularity that is typical of the Mediterranean region. For the purposes of the present study, a relatively wet year (1996/97) was selected, with annual precipitation of 870 mm; this is representative of a hydrogeological situation of high water levels, which is most vulnerable to contamination.

Water circulates within the aquifer in a N-S direction. Transmittivity values are around 10^3 m²/day. Recharge is mainly affected by the infiltration of rainwater and the surface runoff circulating in the Vélez river. Discharge occurs towards the river bed and the sea, including pumped extraction from wells (García Arostegui 1998).

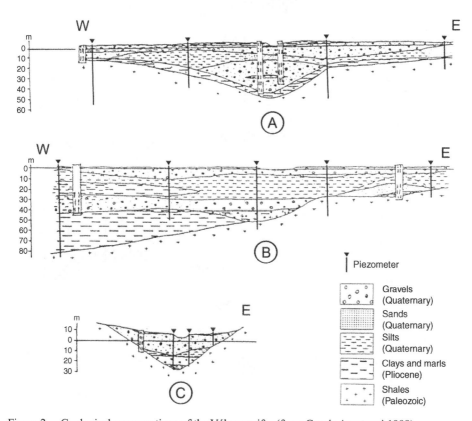

Figure 2.   Geological cross sections of the Vélez aquifer (from García Arostegui 1998).

The Martil-Alila aquifer is constituted of Plio-Quaternary conglomerates, clays, sands and gravels, but lateral changes of facies and granulometry are very frequent, which accentuates the heterogeneity of the aquifer (Figure 3). The substrate is comprised of Tertiary marls and clays, while Palaeozoic metapelitic materials, similar to those described for the Vélez aquifer, outcrop at the borders of the system (Stitou 2002).

Mean annual precipitation is 650 mm and, as with the Vélez aquifer, is characterised by marked annual and interannual irregularity. We selected a year of high precipitation (1996/97) to estimate the piezometric level, a necessary datum for vulnerability mapping.

From the hydrogeological standpoint, the aquifer constitutes a single system, but due to continual facies changes it was not possible to assign a transmittivity value that was representative of the system as a whole. Values range from 100 m²/day to 7,000 m²/day (Stitou 2002). Groundwater flows in a predominantly W-E direction. Recharge is mainly produced by infiltration of rainwater and by runoff from the Martil and Alila rivers. Discharge occurs towards the Martil river bed and the sea, including pumped extraction from wells (Stitou 2002).

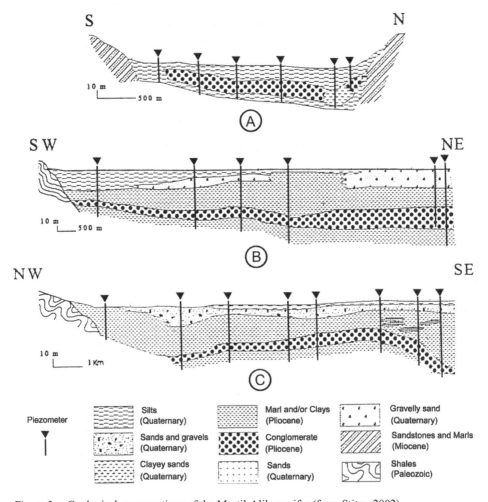

Figure 3.   Geological cross sections of the Martil-Alila aquifer (from Stitou 2002).

2 METHODOLOGY

Vulnerability, or intrinsic vulnerability according to Daly et al. (2002), is the sensitivity of groundwater to contamination, which is determined by the geological, hydrological and hydrogeological features of the aquifer, irrespectively of the type of contaminant and the aquifer media.

In this paper, vulnerability is assessed by the GOD method (Foster & Hirata 1988), which considers the susceptibility of the aquifer to the entry of contaminants from the topographic surface. The following variables are analysed (Figure 4): groundwater occurrence (G), overall lithology of aquifer or aquitard (O) and depth to groundwater table (D).

The D factor depends on recharge and, thus, on climatic conditions. The marked pluviometric irregularity, both annual and interannual, that is typical of the western Mediterranean, requires us to define a specific period when evaluating the vulnerability of an aquifer to contamination. Vulnerability should be analysed for the situation of greatest risk of groundwater contamination, that is, one corresponding to a period of high levels of precipitation, favouring aquifer recharge and the transport of contaminant materials (Goldscheider & Popescu 2004). Aquifer recharge produces a rise in the piezometric level and thus increased vulnerability, due to the reduced transit time between the entry of potential sources of contamination and their contact with the groundwater.

According to Vrba and Civita (1994), the GOD method comprises a mapping overlay based on a factor-scoring system. The variables *Groundwater occurrence* (G) and *Overall lithology of aquifer* (O) are obtained from 1/50,000 scale geological maps. The *Depth to groundwater table* (D) factor is obtained from the difference between the topographic

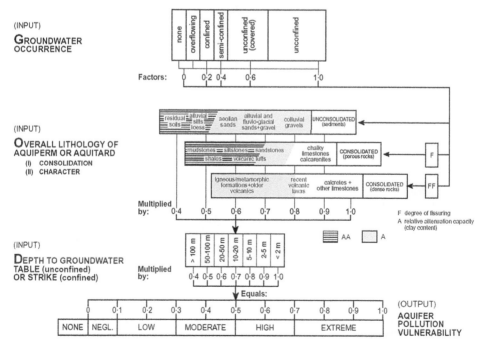

Figure 4. GOD overlay and index system of aquifer vulnerability assessment (from Foster & Hirata 1988).

surface and the piezometric surface, using a Digital Elevation Model (DEM). The super-position of information layers is performed within a Geographic Information System (GIS), which is used to calculate the index and classes of vulnerability.

3   RESULTS

3.1   *Vulnerability Map of Vélez aquifer*

The vulnerability map of the Vélez aquifer shows two well-differentiated sectors, the western half of the aquifer being more vulnerable than the eastern one. In the former, where the aquifer is unconfined, vulnerability is classed as High or Moderate, while in the eastern half, where it is confined, the vulnerability is predominantly Low or Very Low. In no case were any areas classified as presenting Very High vulnerability (Figure 5).

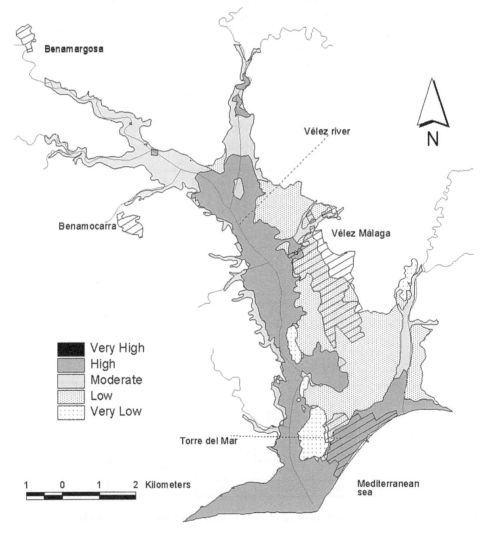

Figure 5.   Vulnerability map of Vélez aquifer by the GOD method.

The zones where vulnerability was classified as High correspond to sectors constituted of alluvial materials, where groundwater depth is less than 10 m. The sectors with Moderate vulnerability are also formed of alluvial materials, but the groundwater depth exceeds 10 m; such sectors are mainly located close to the borders of the aquifer. The sectors with Low vulnerability are found in the Pliocene marly outcrops, where permeability is very low. The sectors with Very Low vulnerability are in the central part of the aquifer, where the impermeable substrate outcrops.

3.2 *Vulnerability Map of Martil-Alila*

The vulnerability map of the Martil-Alila aquifer can be divided into two sectors: in the eastern half, the vulnerability is predominantly High, while it is Low in the western half (Figure 6). In small sectors near the borders of the aquifer, the vulnerability is Moderate or Very Low. No areas of Very High vulnerability were detected.

The western half of the aquifer presented lower levels of vulnerability due to the presence of silts and clays at the surface, which protected the aquifer. The vulnerability in the eastern half, however, is lower because the aquifer is unconfined in this sector.

In the south-easternmost sector, the vulnerability is Very Low due to the presence of Pliocene marls of very low permeability. In the sector where vulnerability is predominantly Low, there is one area, in the north, where vulnerability is High because of the aquifer is unconfined and the unsaturated zone is constituted of gravels and sands.

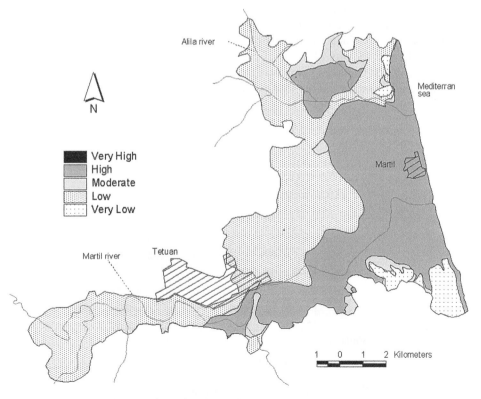

Figure 6.   Vulnerability map of Martil-Alila aquifer by the GOD method.

4 DISCUSSION AND CONCLUSIONS

In the Martil-Alila and Vélez aquifers, the degree of vulnerability and the variations observed are frequently derived from very similar causes. In both aquifers, the vulnerability is mainly a result of the Groundwater occurrence (G) factor. The large differences in vulnerability measured in the study areas are the result of variations in the degree of aquifer confinement. To a lesser degree, variations in vulnerability were observed at the borders of the aquifers, determined by the Depth to groundwater (D) factor; the vulnerability is one degree lower because groundwater depth is greater. Another feature common to both aquifers is the fact that the zones where vulnerability is Very Low correspond to outcrops of the impermeable substrate.

With respect to the proportion of surface area corresponding to each class of vulnerability, the values obtained for the two aquifers were very similar. In the Martil-Alila aquifer, the vulnerability was basically Low or High, while in Vélez, it was High, Low or Moderate (Figure 7). In both aquifers, the percentages of High and Very Low vulnerability were very similar; however, the Vélez aquifer was found to be more vulnerable than Martil-Alila because it presented a higher percentage of surface area with a Moderate degree of vulnerability and a lower percentage with a Low degree of vulnerability.

The vulnerability maps obtained reveal the susceptibility of the Martil-Alila and Vélez aquifers to groundwater contamination derived from the surface. In the case of multilayer aquifers, such as those analyzed in the present study, it is not possible to take a single value for the difference between the topographic surface and the piezometric surface (a parameter which is dependent on the Depth to groundwater, D factor), due to the existence of various strata containing groundwater. Therefore, a necessary prior step in vulnerability mapping for this kind of aquifer is to determine for which stratum the vulnerability is to be evaluated. The vulnerability of the areas in the present study was analyzed for the most superficial materials containing groundwater, due to the impossibility of its evaluation for the whole aquifer by means of the GOD methodology.

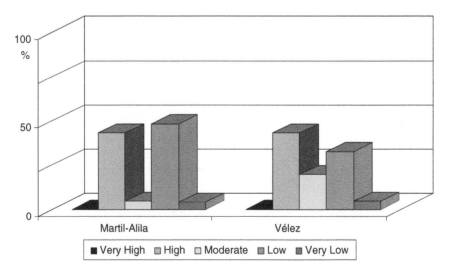

Figure 7. Percentages of surface area by classes of vulnerability.

ACKNOWLEDGEMENTS

This study was carried out as part of the Junta de Andalucía Project A48/02(M) and is a contribution to Research Groups RNM 308 and HUM 776.

REFERENCES

Daly, D., Dassargues, A., Drew, D., Dunne, S., Goldscheider, N., Neale, S., Popescu Ch. & Zwhalen, F. 2002. Main concepts of the "European Approach for (karst) groundwater vulnerability assessment and mapping. *Hydrogeology Journal* 10 (2): 340–345.

Foster, S.S.D. & Hirata, R.C.A. 1988. Groundwater pollution risk assessment: a methodology using available data. WHO-PAHO-CEPIS Technical Report. Lima,-Peru: p. 73.

García Arostegui, J.L. 1998. Estudio hidrogeológico y modelización del acuífero de los ríos Vélez y Benamargosa (Málaga). PhD, Instituo del agua, Univ. Granada. Vol. 1: p. 377.

Goldscheider, N. & Popescu, I.C. 2004. Intrinsic vulnerability: the European Approach. In: *Vulnerability and risk mapping for the protection of carbonate (karst) aquifers.* (Zwahlen ed.) Final report COST Action 620, EUR 20912, European Commission, Directorate General for Research, Luxembourg: 17–21.

Stitou, J.E. 2002. Etude de la salinite des eaux souterraines des aquiferes cotiers Martil-Alila et Smir: Integration des methodes hydrogeochimiques, geophysiques et isotopiques. PhD, Universite Abdelmalek Essaddi of Tetuan: p. 270.

Vrba, J. & Civita, M. 1994. Assessment of groundwater vulnerability. In: *Guidebook on mapping groundwater vulnerability.* (Vrba & Zaporozec eds.), IAH. Verlag Heinz Heise. Vol. 16: 31–49.

Part II   Karst aquifers

CHAPTER 16

DAC: a vulnerability assessing methodology for carbonate aquifers, validated by field and laboratory experiments

F. Celico, E. Petrella & G. Naclerio
Dipartimento di Scienze e Tecnologie per l'Ambiente e il Territorio, Università degli Studi del Molise, Isernia, Italy

ABSTRACT: A hydrogeological and microbiological study has been carried out in order to verify the effectiveness of DAC methodology for "source vulnerability" assessment in fractured and karstified carbonate aquifers. The research has been developed (a) by monitoring the microbial contamination of two springs for three years, and (b) by developing column tests in intact soil blocks, using a collection strain of *Enterococcus faecalis*. The comparison of the vulnerability map with experimental results demonstrated the effectiveness of DAC as a predictor of groundwater microbial contamination in fractured and karstified carbonate aquifers. In particular, it verified the effectiveness of the method to highlight the diversified role of the diffuse infiltration of precipitation through the soil and the bedrock fractures, and the more or less concentrated infiltration of surface water in karst areas. Moreover, the methodology is able to differently analyze the influence of karst systems entirely located within the unsaturated medium or partially developed into the saturated aquifer.

1 INTRODUCTION

Groundwater vulnerability is assessed to identify the areas where the aquifers are characterized by greatest potential for contamination. The vulnerability is often defined utilizing parametric methods (such as DRASTIC; Aller et al. 1987), by assigning point ratings and weights to the individual factors.

The chosen vulnerability methodology should be a function of the type of hydrogeological setting and the type of vulnerability we want to assess. In carbonate massifs of central-southern Italy, the abstraction point are generally located near the springs. In these scenarios, the assessment of "source vulnerability" (Zwalhen 2004) must be preferred to the evaluation of "resource vulnerability". Hence, a methodology which also contains those parameters which are representative of the saturated aquifer, such as "aquifer media" and "hydraulic conductivity", should be chosen.

The method often used to assess vulnerability in carbonate aquifers of central-southern Italy is the DAC (Celico, 1996), the effectiveness of which has been recently verified (Celico & Naclerio 2005; Celico et al. 2006) by comparing the vulnerability values with the results of experiments which were carried out in both the field (through three years of monitoring) and the laboratory (by means of column tests), in agreement with other similar research (Rupert 2001; Perrin et al. 2004). In particular, DAC was shown to be a good predictor of microbial contamination. Hence, according to Zwalhen (2004) and Sinreich et al. (2004), it can be used for "specific vulnerability" assessment. Microbial contaminants have been chosen because in carbonate massifs bacterial pollution is often caused by cattle grazing and manure spreading (Gerba 1985, Pasquarell & Boyer 1995, Boyer & Pasquarell 1999, Celico et al. 1998, 2004a; 2004b). Carbonate aquifers provide the main drinking-water resources of central-southern Italy, supplying an average volume of $4100*10^6 m^3/y$ (Celico et al. 2000).

2 THE DAC METHODOLOGY

The DAC methodology allows the assessment of the pollution potential from direct channeling of surface water into swallow holes and from concentrated recharge in topographically low areas. This method may then be utilized to assess vulnerability of extensively fractured and subordinately karstified carbonate aquifers. DAC comprises two different approaches:

(I) a "classic" DRASTIC approach (Aller et al. 1987) that allows the assessment of the pollution potential from diffuse infiltration of precipitation through the soil and the fractured limestones;

(II) a "new" DRASTIC-based approach (described below) that allows the assessment of the vulnerability from concentrated infiltration of surface water into swallow holes or topographically low areas.

The new approach provides solutions where (A) runoff water infiltrates into swallow holes, with no significant interaction between contaminants and unsaturated rocks and (B) where surface water infiltrates through porous and/or fractured unsaturated media, with a more or less significant interaction between rocks and contaminants. Moreover, in both settings different methodological solutions have been given in order to diversify (1) the cases in which the karst systems are entirely located within the unsaturated medium, and (2) the cases in which the karst systems are partially developed into the saturated aquifer.

The vulnerability assessment in endorheic areas requires the utilization of both the "classic" and the "new" approaches. The vulnerability map must graphically show the results of both approaches. Colors (or different tones of grey) can be used to describe the vulnerability from diffuse infiltration of precipitation, while lines (colored or dark) can represent the pollution potential from concentrated infiltration of runoff water.

The new approach is based on a reinterpretation of some parameters and contains the same weights, ranges and ratings of the "classic" approach. Hence, to determine the relative importance of each parameter, the new methodological approach does not modify the weights through which each factor is evaluated with respect to the other (Table 1). It analyzes some factors in a different way, to adapt their meaning to the specific hydrogeological setting. Seven classes of vulnerability were defined, taking into account the results of the experimental studies (Table 2, Celico et al. 2006).

2.1 *Infiltration of surface water through a porous and/or a fractured unsaturated medium*

Assessing vulnerability from surface water infiltration through a porous and/or a fractured unsaturated medium requires reinterpretation of five of the seven hydrogeologic factors that control the groundwater pollution potential.

Table 1. Assigned weights for DAC features (the weights are the same of the original DRASTIC; Aller et al. 1987).

Hydrogeologic factor	Weights	Pesticide weights
Depth to water	5	5
Net recharge	4	4
Aquifer media	3	3
Soil media	5	2
Topography	3	1
Impact of vadose zone	4	5
Hydraulic conductivity	2	3

Table 2. Degrees of vulnerability in the DAC method.

Vulnerability index	Degree of vulnerability
>200	Extremely high
180–200	Very high
157–179	High
125–156	Medium
95–124	Low
65–94	Very low
<65	Extremely low

"Depth to water" is the thickness of the porous or fractured unsaturated medium through which surface water infiltrates in topographically low areas, determining a more or less significant interaction between contaminants and rocks. "Net recharge" is the amount of runoff water per unit area of land which infiltrates in topographically low areas and percolates to the groundwater. Hence, it represents the increase in contaminants due to the infiltration of treated waste water and/or animal wastes discharged into overlying surface drainages. "Soil medium" is the uppermost portion of the vadose zone characterized by significant biological activity, which crops out in the topographically low areas. It is the soil medium that helps in controlling the migration of contaminants to the groundwater. "Topography" refers to the slope variability of the land surface in endorheic basins. It helps to control the runoff and consequently the amount of contaminants which infiltrates in topographically low areas. Hence, evaluation of the ratings required an inversion of the "classic" DRASTIC relationship to show that the increase of the percent slope produces an increase of the amount of surface water which infiltrates in karst areas, and then a progressive increase of the rating (see Figure 1). "Impact of the vadose zone" determines the attenuation characteristics of the rocks through which surface water infiltrates in topographically low areas.

Different methodological solutions have been given when surface water infiltrates through a porous medium which overlies karst conduits. In this case assessing vulnerability requires a different approach to the parameters "aquifer media" and "hydraulic conductivity". When the karst system (case 1) is entirely located within the unsaturated medium the rating of these two parameters is always set to the maximum (10), because there is not a saturated medium which significantly interact with pollutants. Otherwise, when the karst system (case 2) is partially located within the saturated medium, the factors "aquifer media" and "hydraulic conductivity" are not reinterpreted because they represent the hydrogeologic features of the saturated medium through which the groundwater flows and contaminants are transported.

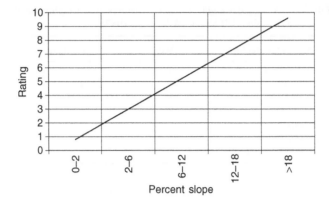

Figure 1.   Graph of ranges and ratings for topography in case of concentrated infiltration of surface water.

2.2   *Infiltration of surface water through swallow holes and karst systems*

Assessing vulnerability from direct channeling of surface water into swallow holes and karst systems requires a different reinterpretation of some factors. The rating of "depth to water", "soil medium" and "impact of the vadose zone" is always set to the maximum (10), because there are no unsaturated media which significantly interact with pollutants, even if the groundwater is hundreds of meters deep. "Net recharge" is the amount of runoff water per unit area of land which infiltrates into swallow holes and karst conduits. As well as when surface water infiltrates through porous and/or fractured media, the increase of the percent slope produces an increase of the amount of surface water which infiltrates into swallow holes. Hence, evaluation of the ratings required an inversion of the "classic" DRASTIC relationship (Figure 1) (Celico 1996; Celico & Naclerio 2005).

Also in this hydrogeological setting, different methodological solutions have been given when surface water infiltrates through a swallow hole into a karst system differently located within the aquifer. In more detail, assessing vulnerability requires a different approach to the parameters "aquifer media" and "hydraulic conductivity". When the karst system (case 1) is entirely located within the unsaturated medium, the rating of these two parameters is always set to the maximum (10), because there is not a saturated medium which significantly inter-act with pollutants. Otherwise, when the karst system (case 2) is partially located within the saturated medium, the factors "aquifer media" and "hydraulic conductivity" are not reinter-preted because they represent the hydrogeologic features of the saturated medium through which the groundwater flows and contaminants are transported.

3   VERIFICATION OF THE EFFECTIVENESS OF THE DAC METHOD

3.1   *Geology, hydrogeology and vulnerability of study area*

The experimental verification of the effectiveness of the DAC method was carried out in a limestone aquifer of southern Italy, which consists predominantly (De Corso et al. 1998) of calcareous deposits (Cretaceous-Oligocene; *Monte Coppe, Coste Chiaravine, Monte Calvello* and *Monaci* Formations) and, subordinately, marly-calcareous (Oligocene-Burdigalian;

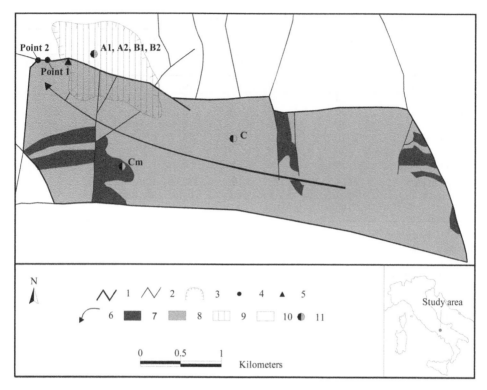

Figure 2.   Vulnerability map of "Acqua dei faggi" carbonate aquifer.
1 – groundwater divide; 2 – faults; 3 – limit of the endorheic area; 4 – spring; 5 – swallow hole; 6 –
groundwater flow direction; 7 – low degree of vulnerability, induced by the diffuse infiltration of
precipitation (Cm); 8 – medium degree of vulnerability, induced by the diffuse infiltration of precipitation (C); 9 – area which produces a (A1) extremely high degree of vulnerability, as a function of
concentrated infiltration of surface water in a swallow hole when the karst system is entirely located
within the unsaturated medium; (A2) very high degree of vulnerability, as a function of concentrated
infiltration of surface water in a swallow hole when the karst system is partially located within the
saturated medium; (B1) very high degree of vulnerability, in case of concentrated infiltration of surface water through a soil superimposed on karst conduits in a topographically low area, when the
karst system is entirely located within the unsaturated medium; (B2) high degree of vulnerability, in
case of concentrated infiltration of surface water through a soil superimposed on karst conduits in
a topographically low area, when the karst system is partially located within the saturated medium;
10 – carbonate deposits located outside the study area; 11 – example locations of Table 3.

Macchiagodena Formation) and calcareous-marly deposits (Langhian-Tortonian, *Cusano* and
Longano Formations). The rocks have very low primary permeability but they are extensively
fractured and subordinately karstified.

The Acqua dei Faggi carbonate massif is bordered by normal faults which have produced low permeability cataclastic zones. The groundwater flows westwards to spring A
(see Figure 2). For several months a year the spring discharged between 1,065 and 1,075
meters a.s.l., depending on the groundwater level fluctuations. The highest outflow (point
1 in Figure 1) is the lowest portion of an explored karst conduit, which is directly interconnected with a swallow hole at 1,091 meters a.s.l. located 350 meters from the spring
Point 1 is active a few times a year. In some cases, it is just fed by the runoff water which

Table 3. Examples of vulnerability assessment in the study area.

	Concentrated infiltration				Diffuse infiltration		
	A1	A2	B1	B2	C	Cm	Weights
Depth to water	10	10	10	10	1	1	5
Net recharge	4	4	4	4	9	9	4
Aquifer media	10	8	10	8	8	8	3
Soil media	10	10	5	5	5	5	2
Topography	10	10	10	10	1	1	1
Impact of vadose zone	10	10	10	10	8	5	5
Hydraulic conductivity	10	4	10	4	4	4	3
Final point sum	206	182	196	172	128	113	
Vulnerability degree	Extremely High	Very High	Very High	High	Medium	Low	

(Case A1: vulnerability from concentrated infiltration of surface water in a swallow hole when the karst system is entirely located within the unsaturated medium; Case A2: vulnerability from concentrated infiltration of surface water in a swallow hole when the karst system is partially located within the saturated medium; Case B1: vulnerability from concentrated infiltration of surface water through a soil superimposed on karst conduits in a topographically low area, when the karst system is entirely located within the unsaturated medium; Case B2: vulnerability from concentrated infiltration of surface water through a soil superimposed on karst conduits in a topographically low area, when the karst system is partially located within the saturated medium; Case C: vulnerability from diffuse infiltration of precipitation through fractured limestone; Case Cm: vulnerability from diffuse infiltration of precipitation through marly-calcareous and calcareous-marly rocks).

infiltrates into the swallow hole. In these cases the karst conduit is entirely located within the unsaturated limestones. In winter the groundwater level episodically reaches the lowest portion of the same karst conduit. In these scenarios, point 1 represents the outflow of the groundwater or both the groundwater and the runoff water which infiltrates into the swallow hole.

The discharge of surface and spring water was monitored at the swallow hole and at both points 1 and 2 (Figure 1). Surface runoff ranged from 0 to 0,017 m^3/s, while the discharge at point 1 ranged from 0 to 0,028 m^3/s. The discharge at point 1 was higher than that measured at the swallow hole when the same point 1 was fed by both surface and groundwater. The whole spring, measured at point 2, had a discharge which ranged from 0,060 to 0,450 m^3/s. The groundwater level was monitored hourly in a piezometer. The comparison between groundwater level fluctuations and discharge at point 2 confirms that the spring is mainly fed by the groundwater which flows within the fracture pattern (Celico et al. 2006), similar to other carbonate aquifers in southern Italy (Celico et al. 2000).

Water level fluctuations in a piezometer in response to pumping from a well show that the limestone aquifer is laterally well connected in the subsurface. A pumping test in which the well was pumped (0.002 m^3/s) for four hours caused 0.51 m of drawdown in the piezometer. Results of this test yielded a transmissivity of 43,2 m^2/d and a storage coefficient of 3×10^{-4}.

The Thornthwaite water-budget method (Thornthwaite and Mather 1957) was used to provide an estimation of net infiltration. The estimation of runoff was obtained by utilizing the experimental results of Boni et al. (1982) related to the surface water monitoring in different catchment areas of Italian carbonate Apennines. The annual average rainfall in the study area is 1240 mm; the annual average evapotranspiration is 500 mm (obtained as a function of an annual average temperature of 9.5°C); the annual average runoff is 110 mm; the annual average net recharge is 630 mm in the exoreic areas, and 740 mm in the endorheic basin, where it represents the sum of both the effective infiltration and the runoff. These values were estimated on the basis of precipitation and temperature data recorded for a period of 80 years (1921–2000). Significant runoff is generally generated during intense rainfall events.

Grazing (a few hundreds cattle) is developed in 56% of the study area. In pasture areas limestones are covered by *Epilepti-Vitric Andosols (Mollic)* (Soil Reference Base; FAO 1988), which is characterized by a profile A/R. The horizon A generally ranges in thickness from 4 to 12 cm.

The aquifer is generally characterized by a medium vulnerability from diffuse infiltration of precipitation (case C in Figure 2 and Table 3). The lowest depth to groundwater table near the springs causes a high vulnerability in a zone too small to be shown on the map. A low vulnerability was calculated where marly-calcareous and calcareous-marly rocks crop out (case Cm in Figure 2 and Table 3).

Moreover, as with the endorheic area, the application of the DAC shows: (1) an extremely high vulnerability from infiltration of surface water into swallow holes, where the karst system is entirely located within the unsaturated medium (case A1 in Figure 2 and Table 3), (2) a very high vulnerability from infiltration of surface water into swallow holes, where the karst system is partially located into the saturated medium (case A2 in Figure 2 and Table 3), (3) a very high vulnerability from concentrated infiltration of runoff water through a soil superimposed on karst conduits in the topographically low area, where the karst system is entirely located within the unsaturated medium (case B1 in Figure 2 and Table 3), and (4) a high vulnerability from concentrated infiltration of runoff water through a soil superimposed on karst conduits in the topographically low area, where the karst system is partially located into the saturated medium (case B2 in Figure 2 and Table 3). Hence, in the endorheic area, the same limestone aquifer shows a significant pollution potential from direct channeling of surface flow into the subsurface and a lower vulnerability from diffuse infiltration of precipitation. Contaminants transported from areas partially located outside the aquifer boundaries (northern part of the endorheic area) can produce pollution of groundwater.

3.2 *Materials and methods*

Surface and groundwater samples were collected weekly or biweekly, from January to July 2001, from December 2002 to May 2003, and from November 2003 to June 2004, in sterile 1,000 ml bottles and transported in a refrigerated box to the laboratory. Filtration processes for bacteriological analyses were made within 2 hours or less from collection. Indicators of microbial contamination were determined using classic methods of water filtration on sterile membranes filter (GN-6 Metricel, pore size 0.45 μm, Pall), with incubation on: (a) mEndo-Agar LES (BBL) for 24 h at 35°C, for total coliforms, (b) m-FC Agar for 24 h at 44.5°C, for fecal coliforms and (c) KF-Streptococcal Agar for 48 h at 35°C, for fecal enterococci.

Two intact soil blocks of *Epilepti-Vitric Andosols (Mollic)* were extracted from the study site, in pasture areas. To minimize the disturbance of samples, sod-covered blocks (181,36 cm square by 11 cm deep) were carved from undisturbed soil directly utilising permeameter cells used for column tests. All blocks were covered in plastic and transported to the laboratory, where the experimental procedure started immediately.

A diffuse interaction between bacteria and soil blocks was obtained by developing column tests in a standard permeameter (Model S248, MaTest S.r.l.) to prevent lateral flow within the gap between soil block and metal cell. One liter of water was poured onto the top of the two soil blocks, at a rate of 60 mm h^{-1}, to simulate the effect of the surface water infiltration through the soil medium, in the topographically low area. The outflow was collected at the bottom. A peristaltic pump was used to sustain a constant flow through the blocks. A solution with 0.001 M CaCl$_2$ was used to prevent dispersion of clays within the soil and column plugging (McMurry et al. 1998).

The interaction between faecal bacteria and soil blocks was analyzed through the utilization of a strain of *Enterococcus faecalis* (ATCC 29212), which is nalidixic acid resistant. No nalidixic acid resistant bacteria were observed in the natural background of soil blocks collected in pasture areas. *E. faecalis* was aerobically grown at 37°C in Luria-Bertani (LB) liquid or solid media (Sambrook et al. 1989). At the beginning of the experiments, 0,75 × 10^9 *E. faecalis* cells (during the exponential growth) were applied at the top of each block in a solution with 0.001 M CaCl$_2$. Soil block drainage was collected in 10 ml sterile plastic tubes beneath the outflowing holes. 200 µl of each water sample and its serial dilutions were plated in triplicate on LB solid medium, supplemented with antibiotic (20 µg/ml of nalidixic acid) and incubated at 37°C. After 24 h the number of *E. faecalis* cells was estimated as colony-forming units (CFU), utilising only the plates where the number of colonies ranged from 30 to 300.

3.3   *Results of the microbiological monitoring*

The microbial contamination of surface water collected in the endorheic area showed a mean concentration of fecal bacteria of some hundreds of CFU/100 ml. (Table 4). High microbial pollution was verified in water samples collected at spring "A" at point 1 (case A1 in Table 4), during intense rainfall events that produced significant runoff and its infiltration in the swallow hole directly interconnected with the spring. In this case the spring

Table 4.   Main results of microbiological experiments.

	Mean concentration of fecal enterococci (CFU/100 ml)	Number of samples
Surface water	291	9
Case A1	271	6
Case A2	48	3
Case B1 (simulated)	29	6
Case B2 (simulated)	5	3
Case C	0	36

(Case A1: concentrated infiltration of surface water in a swallow hole when the karst system is entirely located within the unsaturated medium; Case A2: concentrated infiltration of surface water in a swallow hole when the karst system is partially located within the saturated medium; Case B1: concentrated infiltration of surface water through a soil superimposed on karst conduits in a topographically low area, when the karst system is entirely located within the unsaturated medium; Case B2: concentrated infiltration of surface water through a soil superimposed on karst conduits in a topographically low area, when the karst system is partially located within the saturated medium; Case C: diffuse infiltration of precipitation).

has the same degree of contamination that was observed in surface water samples, where the karst system is entirely located within the unsaturated medium. On the other hand, when the karst system significantly interacted with the groundwater, the water samples collected at point 1 of spring A showed 48 CFU/100 ml of fecal bacteria, that is to say a degree of contamination significantly lower than that detected in surface water samples. Hence, in spring water samples it was detected a microbial contamination which was the result of interaction between surface runoff and groundwater.

Water samples collected at spring "A" at point 2 (case C in Table 4) did not show, on average, a significant microbial pollution during precipitation which did not produce the infiltration of surface runoff in the swallow hole. Hence, the diffuse infiltration of precipitation did not allow a significant contamination due to the interaction between microorganism and both the soil and the aquifer.

3.4 *Results of the column tests*

The concentrated infiltration of surface water through a soil medium superimposed on karst conduits in a topographically low area should produce a microbial contamination of spring water lower than that caused by the infiltration in the swallow hole. This is due to the interaction between microorganisms and soil, before their migration into the karst system. In order to quantify the effect of this phenomenon, column tests in intact soil blocks collected in pasture areas have been carried out.

The total number of bacteria eluted after 1 liter of water represented from the 7% to the 12% of the inoculated *E. faecalis* cells in the blocks, confirming that the analyzed soil medium is characterized by a significant retention of bacteria. The significant storage capacity of microbes was already verified utilizing different soils and microorganisms (i.e. Gannon et al. 1991, Trevors et al. 1990). Hence, the filtration of contaminated surface water through the soil in a topographically low area will produce a significant decrease (about 90% in study site) of the amount of bacteria transported to the groundwater, during each infiltration event. Taking the microbial contamination of runoff into consideration (291 CFU/100 ml; surface water in Table 4), the mean value of fecal enterococci transported below the soil bottom in the karst plane should be 29 CFU/100 ml (10%; case B1 in Table 4) when the karst system is entirely located within the unsaturated medium. In contrast, in the case of karst systems partially located within the saturated medium, a lower contamination is expected, due to the interaction between bacteria and groundwater. In more detail, taking into account the results of both the field (case A2) and the lab experiments, the migration of microorganisms through the soil before their transport into the karst system should produce a microbial contamination of a few UFC/100 ml (case B2 in Table 4).

4 COMPARISON OF VULNERABILITY MAP WITH GROUNDWATER QUALITY DATA

Groundwater quality data were compared with vulnerability categories produced by both DRASTIC and DAC to verify the effectiveness of the latest as a predictor of microbial contamination in extensively fractured and subordinately karstified carbonate aquifers. The vulnerability map, obtainable through the application of DRASTIC, has a poor correlation with microbial contamination data. In fact, it calculated a vulnerability generally ranging

Table 5.   Comparison of vulnerability category derived using the DAC methodology and microbial data.

Case	Transport modalities	Vulnerability degree	Mean concentration of fecal enterococci [CFU/100 ml]
A1	Concentrated infiltration of surface water in a swallow hole (karst system entirely located within the unsaturated medium)	Extremely high	271
A2	Concentrated infiltration of surface water in a swallow hole (karst system partially located within the saturated medium	Very high	48
B1	Concentrated infiltration of surface water through soil (karst system entirely located within the unsaturated medium; simulated)	Very High	29
B2	Concentrated infiltration of surface water through soil (karst system partially located within the saturated medium; simulated)	High	5
C	Diffuse infiltration of precipitations	Medium-Low	0

from low to medium while significant microbial contamination was often detected at springs (up to several hundreds CFU/100 ml).

The vulnerability map developed by the DAC methodology had a good correlation with groundwater quality data. In fact (Table 5): in case A1 the concentrated infiltration of surface water into the swallow hole, when the karst system is entirely located within the unsaturated medium, caused an extremely high microbial contamination and determined an extremely high pollution potential; in case A2 the concentrated infiltration of surface water into the swallow hole, when the karst system is partially located within the saturated medium, produced a very high microbial contamination and determined a very high pollution potential; in case B1 the concentrated infiltration of surface water through a soil medium superimposed on karst conduits in a topographically low area, when the karst system is entirely located within the unsaturated medium (simulated by means of column tests), generated a very high microbial pollution and determined a very high vulnerability; in case B2 the concentrated infiltration of surface water through a soil medium superimposed on karst conduits (simulated by means of column tests) revealed a significant microbial contamination and determined a high pollution potential; in case C the diffuse infiltration of precipitations through fractured limestone did not produce any microbial contamination of groundwater and generally determined a medium-low vulnerability. The results of this correlation demonstrate that the vulnerability map produced by DAC is effective in showing the higher influence of concentrated infiltration of surface water in karst areas on groundwater pollution potential.

5   DISCUSSION AND CONCLUSIONS

Both field and laboratory experiments demonstrated the effectiveness of the DAC method for vulnerability assessment in carbonate aquifers. The verification of its effectiveness was obtained by comparing results of vulnerability assessment with groundwater quality data. Due to the type of land use generally and widely developed within carbonate aquifers of central-southern Italy (grazing and not intensive agriculture with use of manure), the verification was developed by carrying out microbiological experiments. In extensively fractured and subordinately karstified carbonate aquifers, the most significant weakness of DRASTIC is the wrong interpretation of the hydrogeologic factors where both diffuse infiltration of precipitation and concentrated infiltration of surface runoff can transport contaminants toward the groundwater. In endorheic areas the higher contamination of groundwater produced by the concentrated infiltration of runoff water into swallow holes and/or topographically low zones can be predicted through reinterpretation of the parameters, as defined in the DAC approach delineated here. Hence, a complex system such as a carbonate aquifer, fractured and subordinately karstified, was simplified in a schematic set of coexisting cases of vulnerability. For each one a methodological solution was found, and for each solution a specific microbiological investigation was carried out, in order to experimentally verify its effectiveness and the reliability of the whole method.

The same methodological approach can be also used in other types of aquifers, to assess vulnerability induced by interaction between surface and groundwater.

Groundwater vulnerability maps developed by DAC will better identify areas of greatest potential for microbial contamination in carbonate aquifers and then they can be used to focus pollution prevention programs on areas of greatest concern, developing a sustainable land use, with emphasis on grazing and manure spreading.

ACKNOWLEDGEMENTS

This project was founded by the European Union (KATER and KATER II Research Programs, INTERREG IIC, CADSES, and INTERREG IIIB, CADSES, respectively) and by the Research National Council of Italy (CNRG00D43F).

REFERENCES

Aller, L., Bennet, T., Leher, J.H., Petty, R.J. & Hackett, G. 1987. DRASTIC: a standardized system for evaluating ground water pollution potential using hydrogeological settings. EPA /600/2-85/018, Washington DC, USA.

Boni, C., Bono, P., & Capelli, G. 1982. Valutazione Quantitativa dell'Infiltrazione Efficace in un Bacino Carsico dell'Italia Centrale. Confronto con Analoghi Bacini Rappresentativi di Diversa Litologia. *Geologia Applicata e Idrogeologia* 17, No 2: 437–452.

Boyer, D.G. & Pasquarell, G.C. 1999. Agricultural Land Use Impacts on Bacterial Water Quality in a Karst Groundwater Aquifer. *Journal of the American Water Resources Association* 35, No 2: 291–300.

Celico, F. 1996. Vulnerabilità all'Inquinamento degli Acquiferi e delle Risorse Idriche Sotterranee in Realtà Idrogeologiche Complesse: i Metodi DAC e VIR. *Quaderni di Geologia Applicata* 1: 93–116

Celico, F., Celico, P., De Vita, P. & Piscopo, V. 2000. Groundwater flow and protection in the Southern Apennines (Italy). *Hydrogeology* 4: 39–47.

Celico, F., Esposito, L. & Piscopo, V. 1998. Interaction of Surface-Groundwater in Fragile Hydrogeological Settings in Southern Italy. In: *Proc. 28th* IAH Congress *"Gambling with Groundwater – Physical, Chemical, and Biological Aspects of Aquifer-Stream Relations"*, *Las Vegas* (USA), September 27–October 2: 475–485.

Celico, F., Musilli, L. & Naclerio, G. 2004b. The impacts of pasture and manure spreading on microbial groundwater quality in carbonate aquifers. *Environmental Geology* 46, No 2: 233–236.

Celico, F. & Naclerio, G. 2005. Verification of a DRASTIC-based method for limestone aquifers. *Water International* 30, No 4: 530–537.

Celico, F., Petrella, M. & Naclerio, G. 2006. Updating of a DRASTIC-based method for specific vulnerability assessment in carbonate aquifers. *Water International* (in press).

Celico, F., Varcamonti, M., Guida, M. & Naclerio, G. 2004a. Influence of precipitation and soil on transport of fecal enterococci in limestone aquifers. *Applied and Environmental Microbiology* 70, No 5: 2843–2847.

De Corso, S., Scrocca, D. & Tozzi, M. 1998. Geologia dell'Anticlinale del Matese e Implicazioni per la Tettonica dell'Appennino Molisano. *Bollettino Società Geologica Italiana* 117: 419–441.

FAO. 1988. Soil Map of the World. Revised Legend. Reprinted with corrections. *World Soil Resources Report* 60. FAO, Rome.

Gannon, J.T., Mingelgrin, U. & Alexander, M. 1991. Bacterial Transport through Homogeneous Soil. *Soil Biology Biochemistry* 23: 1155–1160.

Gerba, C.P. 1985. Microbial Contamination of the Subsurface. In: *Groundwater Quality*, ed. C.H. Ward et al. John Wiley & Sons. New York: 54–67.

McMurry, S.W., Coyne, M.S. & Perfect, E. 1998. Fecal Coliforms Transport through Intact Soil Blocks Amended with Poultry Manure. *Journal Environmental Quality* 27: 86–92.

Pasquarell, G.C. & Boyer, D.G. 1995. Agricultural Impacts on Bacterial Water Quality in Karst Groundwater. *Journal Environmental Quality* 24: 959–969.

Perrin, J., Pochon, A., Jeannin, P.Y. & Zwalhen, F. 2004. Vulnerability assessment in karstic areas: validation by field experiments. *Environmental Geology* 46, No 2: 237–245.

Rupert, M.G. 2001. Calibration of the DRASTIC Ground Water Vulnerability Mapping Method. *Ground Water* 39, No 4: 625–630.

Sambrook, J., Fritsch, E.F. & Maniatis, T. 1989. Molecular Cloning. A Laboratory Manual. 2nd Edition. Cold Spring Harbour Laboratory Press, Cold Spring Harbour, NY.

Sinreich, M., Kozel, R., Mudry, J. & Kralik, M. 2004. A European perspective on specific groundwater vulnerability. In: *Proc. IAH International Conference on "Groundwater Vulnerability Assessment and Mapping"*, Ustron (Poland), June 15–18: 127–128.

Thornthwaite, C.W. & Mather, J.R. 1957. Instructions and Tables for Computing Potential Evapotranspiration and the Water Balance. 5th printing. C.W. Thornthwaite Associates, Laboratory of Climatology, Elmer, NJ, USA, 10(3).

Trevors, J.T., Van Elsas, J.D., Van Overbeek, L.S. & Starodub, M. 1990. Transport of a Genetically Engineered Pseudomonas fluorescens Strain through a Soil Microcosm. *Applied Environmental Microbiology* 56: 401–408.

Zwahlen, F. (ed.) 2004. Vulnerability and risk mapping for the protection of carbonate (karst) aquifers. COST Action 620 Final Report. Office for Official Publications of the European Communities, Luxembourg, XVIII: p. 297.

CHAPTER 17

VURAAS – vulnerability and risk assessment for Alpine aquifer systems

G. Cichocki & Ht. Zojer
Aquaterra Consultants & Experts, Graz, Austria

ABSTRACT: The concept VURAAS was developed in the high Alpine karst region of Nassfeld/Austria. One main result, represented in the vulnerability map showing 5 different areas with distinguishable vulnerability, is obtained by describing the input, infiltration and exfiltration. A second map, the map of hazards shows all kind of hazards and the probability harmfulness of contaminants expressed by the hazard index (HI). The combination of both maps lead to a risk map which shows areas at varying degrees of potential groundwater contamination risks. The risk map is a useful tool for land planners, enabling respect of vulnerable landscapes on the one hand but allows making use of less vulnerable parts of the area for tourism expansion on the other.

1 INTRODUCTION

The constantly growing tourism in high Alpine karst regions presents a growing threat to Alpine groundwater. The use of land for tourism needs to be reconciled with the sustainable protection of groundwater.

The European COST Action 620 Project – "*Vulnerability and risk mapping for the protection of carbonate (karst) aquifers*" – was initiated in 1997, following the ending of COST Action 65 (1995). Delegates from 15 European countries tried to develop a flexible approach for vulnerability mapping and risk assessment motivated by the demands of the European Water Framework Directive (2000). As part of this work, "VURAAS" (**VU**lnerability and **R**isk assessment for **A**lpine **A**quifer **S**ystems) concept was developed in Austria for the COST Action 620 type-site representing "(*high*) *Alpine karst areas*".

The aim of VURASS is to produce a risk map that identifies more vulnerable areas that require groundwater protection and less vulnerable areas more suited to development, such as tourism. The risk map is therefore a useful tool for decision makers and land-use planners. This paper gives a general overview of the VURAAS concept.

2 AN OVERVIEW OF THE CONCEPT OF VURAAS

By combining a *vulnerability map and a hazard map*, VURAAS produces a *risk map* of potential groundwater risks to contamination. A number of individual investigations have contributed to the VURAAS maps, which are outlined below:

- evaluation of artificial rain tests with sprinkler construction on skiing slopes and alpine pastures
- calculation of runoff coefficients for different land uses
- influence of man made snow on the runoff in surface water channels
- analyses of flood events and recession curves
- analyses of hydrochemistry and isotope hydrology for the characterisation and delineation of individual catchment areas.

These specific investigations are detailed in Cichocki (2003) and Zojer (2003).

The vulnerability mapping is based on three core parameters: the input, the infiltration and the exfiltration, which comprise a number of elements (Figure 1).

The theory of the VURAAS concept is shown in the Figure 1.

The three core factors of the input, the infiltration and the exfiltration are described by parameters (see section 3.3). The basic for the classification of each parameter is a point-count-system algorithm (Cichocki 2003), (Zojer 2003). The parameters for the core factor and the procedure of the application of VURAAS for mapping intrinsic vulnerability are explained in the section 3.3.

2.1   Hazard mapping

The *hazards map* is based on the work and recommendations of COST Action 620 (2002). The map shows all of the potential hazards, which are expressed as a point, line or area. The probability of an event occurring combined with the magnitude of the contaminant are also expressed on the map by a "hazard index (HI)", which ranges from 1 to 5. The application of hazard mapping in the test site is explained in section 3.4.

2.2   Risk mapping

Finally a map of intrinsic vulnerability and a map of hazards were combined to create a risk map. The result of the risk map in the test site Nassfeld is shown in section 3.5.

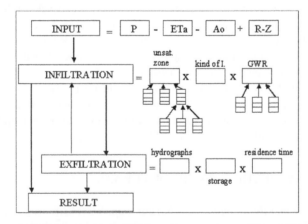

Figure 1.   Scheme of VURAAS for mapping intrinsic vulnerability. (Small boxes within the scheme represent sub-parameter describing the parameter of the unsaturated zone and groundwater recharge (GWR) which are explained in section 3.3.2).

3 APPLICATION OF VURAAS IN THE NASSFELD TESTSITE

3.1 *General*

The austrian approach VURAAS for mapping groundwater vulnerability was applied in an Alpine karst region in the southern Alps of Austria. The test site (about 8 km^2) is located in the "skiing region of Nassfeld" on the Trogkofel Mountain in the Carnian Alps. The altitude within the study area ranges from 1120 m above sea level (a.s.l.) to 2280 m a.s.l. and the average slope gradient is approximately 30°. The mean annual precipitation is c.2260 mm and the mean annual temperature is 3.6°C. On average, snow cover lasts 193 days a year.

3.2 *Geological characteristics of the investigated area*

The geological formations predominantly consist of Permian limestone, which are underlain by schist and sandstones. The carbonates are fractured and slightly karstified. In some areas, the hard rock formations are overlain by rock debris or by moraines. Many springs are located at the contact of the limestone with the "low permeable rocks" at altitudes between 1150 m and 1900 m. A number of these karst springs are used for public water supply ranging from supplying alpine farmer houses, to a tourist centres (Sonnleitn Village) and in the Gail River Valley.

3.3 *Mapping the vulnerability*

3.3.1 *Description of the factor input*
The parameter *input* was determined by using the water balance equation where:

$$Input = P - ETa - Ao + (R \pm \Delta S)$$

with
Input.................amount of recharge [mm/a]
P.......................precipitation [mm/a]
ETa...................actual evapotranspiration [mm/a]
Ao......................surface and subsurface runoff
(R $\pm$ ΔS) ....Retention / drainage and differences in the water storage in the unsaturated zone.
The precipitation [P] and the evapotranspiration [ETa] are calculated using each gradient of altitude together with the digital elevation model. The surface and subsurface flow [Ao] are classified for each landuse in the individual catchment areas. Therefore the runoff coefficients of rain tests and out of literature are taken into consideration.

The parameter R $\pm$ ΔS has low influence to the system and can be neglected.

3.3.2 *Description of the infiltration factor*
In order to determine the infiltration, the material overlying the bedrock must be assessed. The assessment of the overlying layers is based on the PI method (Goldscheider et al. 2000). The type and thickness of the soil were obtained from 60 hand auger holes (7–8 holes per km^2). A representative profile for each type of soil was described and soil material was analysed for the calculation of the effective field capacity. Seven types of unconsolidated material can be distinguished in the testsite Nassfeld. The subsoil thickness was

estimated using geological cross sections and the type of subsoil was mapped in the field. The geology was classified in 5 different categories. The thickness of the non-karst and unsaturated karst bedrock, was estimated using cross sections. The geological structures were determined by a combination of field mapping and interpreting aerial images.

Three infiltration types – point, line and diffuse – were mapped in the field.

In the VURAAS methodology, the groundwater recharge (GWR) takes account of the surface water retention and the density of springs – both recorded during field mapping – and the water storage capacity, which is determined by isotope and hydrochemical analyses (Zojer 2003).

Description of the exfiltration factor – E

For the VURAAS method, the entire catchment area was characterised by analysis of the discharge (exfiltration). If the available data within the catchment are too limited, the discharge component is not included in the overall assessment.

Long term measurements for two hydrological years of the springs' discharge, hydrochemistry, electric conductivity, water temperature and environmental isotopes (^{2}H, ^{18}O) were undertaken and assessed to understand the dynamics and storage capacity of the entire aquifer. Thus, the fluctuation of the hydrograph, the water temperature and conductivity is expressed by statistical evaluation using the variation coefficient.

Recession curves were evaluated to determine the storage capacity, which is expressed by retention coefficient α (Maillet 1905). The mean residence time of water of the springs was determined using tritium (^{3}H) analysis.

A GIS-based production of the intrinsic vulnerability map (Figure 2) in the study area was performed. Therefore GIS layers showing the results of each description of the parameter were overlapped with GIS techniques following the weighting and rating system of VURAAS.

3.4   *Hazard mapping using VURAAS*

3.4.1   *Overview*

Land-use in the Nassfeld test-site is as follow: 46% is covered by forests, 25% by low vegetation ground 13% by alpine pasture and 6% by outcropping rocks, 4% by ski courses, with the remainder recorded as gravel, pasture or low vegetation (Cichocki 2003).

The eastern part of the study area is characterized by winter tourism infrastructure with several ski runways on the Trogkofel and Zweikofel Mountains and additional runways recently built. In winter, the Alpine farm houses are used as ski accommodation.

Description of hazards

The types of hazards within the Nassfeld test site were identified and evaluated according to the approach outlined by COST Action 620 (2002). The hazards were mapped in the summer of 2002, at a mapping scale is 1:10.000.

Fifteen different kinds of point, line and areal hazards were distinguished: The "point" hazards included septic tanks, treatment plants, storage fuel tanks, cesspools, manure heap, animal barns and car parks. Such hazards were identified by local knowledge and field mapping and were generally found around alpine farm houses and ski lifts. Additional information about livestock densities and barn sizes were available from the land-register "Almkataster Nr.1053 and 1050".

Linear hazards, such as forest tracks, were identified from topographic maps. From November to May, these tracks are covered by natural and mechanically-made snow and

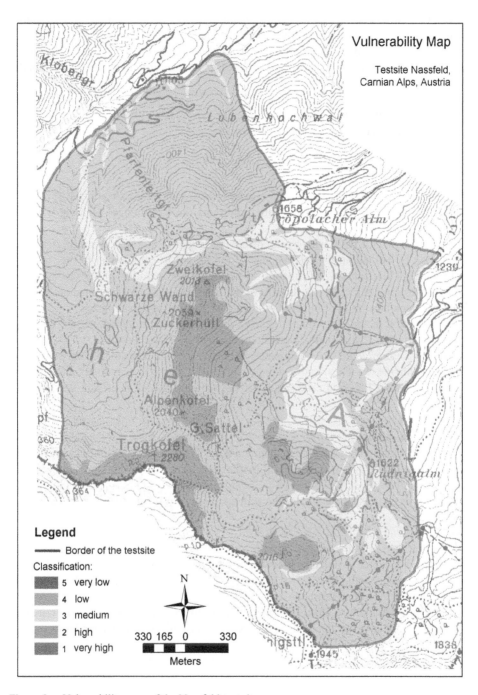

Figure 2.   Vulnerability map of the Nassfeld test site.

are used as ski courses. The existing and planned waste water pipelines were available from a number of environmental protection assessments.

Area hazards comprised the ski courses, which were identified on the line hazards map. Fertilisers (mainly organic) are spread on the ski courses during the summer months including N and P.

3.4.2   Determination of hazard index (HI)

The *hazard index* (HI) is expressed by the formula:

$$HI = H * Qn * Rf.$$

The weighting factor (H) for the 15 different hazards was obtained using the list of hazards of COST Action 620 (2002). This value represents the harmfulness of hazards to the groundwater.

The ranking factor (Q_n), which can correct the H values to a maximum of 20%, was determined by evaluating the relative size of the hazard in comparison to the "average occurrence". Therefore any information of the quality or quantity of substances of the same category was relevant, e.g. the size of animal barns, cesspools and manure heap or the living stock units or the estimated frequency of cars on the forest tracks.

The reduction factor (R_f), which describes the probability of contamination, was estimated empirically. Theoretically the reduction factor may range from 1 to 0.

The "low" *hazard index* class dominates within the test site. It has to be emphasised that hazards within the test site are not continuously present during a year. For example, fertilizers are spread in June and can remain until October. Livestock is present nearly in the same period.

3.4.3   Graphical interpretation

The graphical interpretation helps to distinguish easily between point, linear and polygon hazards. In a GIS-program standardised marker sets were used, which are similar to the proposed symbols of COST Action 620. The signature and symbols are printed in the legend below for giving an example for the application in Nassfeld test site (see Figure 3).

3.4.4   Usefulness of hazard map for test site

The hazard map was mapped at a scale of 1:10.000 and printed at a smaller scale (1:20.000). The larger scale allows for interpretation of the linear and area hazards but a more detailed map would be required for decision making purposes around the point hazards i.e. around the alpine farm houses.

The hazards are likely to change in the future. Specifically, waste water will be removed by pipelines instead of the treatment plants i.e. some hazards may disappear. However, the development of additional ski courses will introduce further ski transport systems (ski lifts, cable cars) and cesspools, which will increase the number of hazards.

3.5   Discussion and conclusions

The risk map is created by combining the vulnerability map and the hazard map. The risk map is a useful tool for land planer and decision maker to respect the most risky parts of an area and to allow any kind of infrastructural and tourist activities in the areas with lower risks.

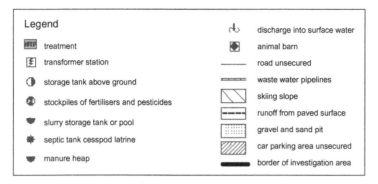

Figure 3.   Legend of the marker sets used in the test site Nassfeld.

REFERENCES

Cichocki, G. 2003. Wassertransport als Grundlage für eine Vulnerabilitätsbewertung und Risikoanalyse in Karbonatgesteinen. *Testgebiet Nassfeld, Karnische Alpen*. Diss. an der TU Graz.

COST Action 620-Final Report 1st draft, 2002. Vulnerability and risk mapping for the protection of carbonate (karst) aquifers. 1st draft of the complete final report.

COST 65 1995. Hydrogeological aspects of groundwater protection in karstic areas, Final report (COST Action 65). European Commission, Directorat-General XII Science, Research and Development, Report EUR 16547 EN; Brussels, Luxemburg: p. 446.

European Water Directive 2000. Directive 2000/60/EC of the European Parliament and of the Council of 23. Oct. 2000 establishing a framework for community action in the field of water policy.

Goldscheider, N., Klute, M., Sturm, S. & Hötzl, H. 2000. The PI method – a GIS-based approach to mapping groundwater vulnerability with special consideration of Karst aquifers. *Z. angew. Geol.*, 46 (2000) 3, Hannover: 157–166.

Maillet, E. 1905. Mécanique et physique du globe. Essais d'hydraulique souterraine et fluviale.-218 S. Paris.

Zojer, Ht. 2003. Stofftransport als Grundlage für eine Vulnerabilitätsbewertung und Risikoanalyse in Karbonatgesteinen. *Testgebiet Nassfeld, Karnische Alpen*. Diss. an der TU Graz.

CHAPTER 18

Groundwater circulation in two transboundary carbonate aquifers of Albania; their vulnerability and protection

R. Eftimi,[1] S. Amataj[2] & J. Zoto[2]
[1] *ITA Consult, Tirana, Albania*
[2] *Institute of Nuclear Physics, Tirana, Albania*

ABSTRACT: The groundwater circulation of transboundary aquifers of the Mali Gjere massif and of the Prespa-Ohrid Lakes is described. The Blue Eye Spring issues from the Mali Gjere karst massif with the mean discharge of $18.4\,m^3/s$. Using environmental tracer methods (isotope and hydrochemical) we have determined that the Blue Eye spring is recharged to the extent of 30 to 35% by the gravely aquifer of Drinos River valley. The high sulphate content of the Drinos River replenishing the gravely aquifer is responsible also for increased sulphate content of the Blue Eye Spring. A karst massif separates the Prespa and Ohrid Lakes and the Prespa Lake is 155 m higher than the Ohrid Lake. Using environmental isotopes we determined that from 38 to 53% of the water from some big karst springs issuing in Ohrid Lake is recharged by the Prespa Lake through the Mali Thate-Galichica karst massif. Water tracing has established karst water flow velocities up to 3200 m/h.

1 INTRODUCTION

The approaches adopted for the vulnerability assessment range from the empirical classification of key properties to the process-based simulation models. This is the reason that the maps vary according to physiography of the study area, purpose of the study, and quantity and quality of data (Civita 1994, Vrba & Zaporozec 1994, Gogu & Dassargues 2000). One of the latest methodologies proposed for karst-groundwater-vulnerability assessment and mapping is based on the so-called "European approach" (Daly et al. 2002). The first step towards the application of the "European approach" for mapping groundwater vulnerability is to asses the circulation of groundwater of carbonate aquifers leading to the construction of a conceptual hydraulic network. Care should be taken in application of the models (White 1969, Shuster & White 1971) because is very difficult to know the proportions of diffuse and of conduit flow in a carbonate aquifer (Atkinson 1976, Motyka 1998).

In this paper we attempt to clarify the groundwater circulation in two very important transboundary carbonate aquifers of Albania emphasizing the role of the environmental (isotope and hydrochemical) and the artificial tracer methods of investigation.

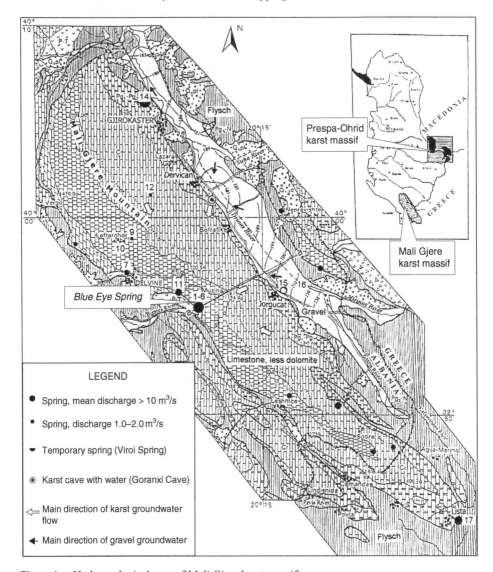

Figure 1.   Hydrogeological map of Mali Gjere karst massif.

2   MALI GJERE KARST MASSIF

2.1   *Geological and hydrogeological characteristics of study area*

The Mali Gjere karst massif is located in south Albania on the border with Greece (Figure 1); its total surface area is 440 km², mostly located in Albanian territory (54 km² in Greek territory). The highest point of the massif is at 1,798 m a.s.l., while the mean altitude is about 900 m a.s.l. The crest of the Mali Gjere Mountain is the natural water divide between the Drinos River basin located on the east, and the Bistrica River basin located on the west.

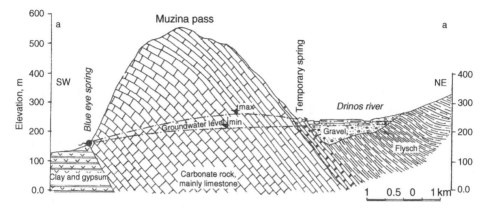

Figure 2.   Cross-section a-a in Mali Gjere karst massif.

Some sulphate springs recharge the Drinos River in Greek territory; the biggest of them is Rogozi Spring with the mean discharge of about $0.5\,m^3/s$ and with the sulphate ion concentration of about 700 mg/l.

The geology of the study area and the sampling locations are shown in Figure 1. The Mali Gjere Mountain is an anticline dipping to the east with 25–30°, while the structure is overthrown to the west (Figure 2). The major units constituting the study area are the carbonate formations, like the Triassic dolomites, the Jurassic limestone with siliceous rocks and the Cretaceous and Palaeogene stratified limestone. The carbonate rocks are surrounded mainly by the Palaeogene and Noegen flysch formation, while the Permo-Triassic clayey-gypsum has a small outcrop in the western side of the Mali Gjere Mountain. In the central-eastern side of the karst massif, in the Jorgut-Dervican area, along a length of 6.5 km, the carbonate rocks contact the Quaternary gravel deposits of the Drinos River valley (AHS 1985).

The correlation between the mean yearly precipitation, P (mm), and the altitude of the recording site, E (m), is determined by monitoring the precipitation during 1960–1985 at four stations at various altitudes within the study area. Results are expressed by Equation (1), for which the correlation coefficient (r) is 0.83:

$$P = 1698 + 0.625\,E \tag{1}$$

Using the method described by Kessler (1967) the effective infiltration of the mean precipitation recharging the carbonate aquifer of Mali Gjere Mountain is estimated at 52%, equal to 1175 mm/year ($517*10^6\,m^3$/year or $16.4\,m^3/s$). The remaining 48% of the mean precipitation represents the evapotranspiration (about 38%) and surface runoff (about 10%). Most of the karst water drains to the western side of the Mali Gjere massif where the Blue Eye Spring (mean discharge $18.4\,m^3/s$) issues at an elevation of about 45 m lower than that of the Drinos Valley. Also some springs, each with a discharge of less than $0.1\,m^3/s$, issue from this side of the massif. The biggest spring of the eastern side of the massif is the Viroi ephemeral spring (maximal discharge $35\,m^3/s$). The total discharge of all the springs of Mali Gjere karst massif results about $743*10^6\,m^3$/year, ($23.6\,m^3/s$). By the balance calculations results that the total discharge of the springs of studied karst massif is about 30–35%

bigger than the calculated mean efficient precipitation of the massif, which corresponds to a water quantity of about $226*10^6 \, m^3/year$ ($7.17 \, m^3/s$).

2.2   *Environmental tracer approach[1]*

Environmental isotopes are most useful in problems related to the origin of the water and the dynamics of water systems. Based on the altitude effect, isotopes may be used to identify the potential source areas of recharge (Bradley at al. 1972, IAEA 1981, Payne et al.1978). Environmental isotope techniques are used here together with the hydrological and hydrochemical methods to verify the partial replenishment of the karst water resources of Mali Gjere karst massif by the gravely aquifer of the Drinos River.

The formulation of this hypothesis takes into consideration a good hydraulic connection between the gravely and karst aquifers, as well as the natural groundwater slope to the Blue Eye Spring, (Figure 2). Sample collection began in January 1988 and continued until December 1999, and some sporadic sampling was done during 1996. The samples were analyzed for oxygen-18 and deuterium. The sampling program included 6 springs located at altitudes of 150–1000 m (Blue Eye Spring is sampled at six main outlets), one borehole in Drinos Valley and the Drinos River. The results of the isotope measurements performed by Isotope Hydrology Laboratory of IAEA, and some hydrochemical data are presented in Table 1. The analytical errors are 0.1‰ for $\delta^{18}O$ and 1.0 ‰ for δD.

2.3   *Discussion and interpretation of isotope data*

The isotopic composition of all the six outlets of the Blue Eye Spring (no 1–6) is very homogenous; the standard deviation of mean values for each outlet varies within 0.02–0.08 for $\delta^{18}O$ and within 0.6–1.1 for δD. This is an indication of the good mixing of the karst groundwater of Mali Gjere massif. The presented in Table 1 isotope values of Blue Eye Spring are the weighted ones of the mean values of six sampled outlets of this spring.

The correlation function between the mean $\delta^{18}O$ and δD values of the sampling points showed in Figure 3 results in two equations (2) and (3):

$$\delta D = 7.46 \, \delta^{18}O + 14.51 \tag{2}$$

$$\delta D = 2.28 \, \delta^{18}O - 26.58 \tag{3}$$

Equation (2) indicates an average slope of 7.46, which is close to the slope 8 of the meteoric water line. The δ-values obtained in Blue Eye Spring, in borehole (no 15) and in Drinos River (nr 16) scatter around a mixing line which slope is m = 2.28.

The recharged by the Drinos River the gravely aquifer groundwater is more affected by the relative δD and ^{18}O enrichment. It seams that the mean recharge area of the Drinos River catchments has lower elevation that this of the Mali Gjere karst massif. However we do not know the contribution of the big sulphate springs issuing in Greek territory to the isotopic composition of the Drinos River.

The values of the intercept of the mixing line with the meteoric water line: $\delta^{18}O = -7.9‰$ and $\delta D = -44.3$ ‰ coincide with the isotopic composition of the Viroi ephemeral spring

---

[1] Prof. J.G. Zötel and Prof. H. Zojer have given much useful advice in the early stage of this investigation.

Table 1. Analyses for environmental isotopes in surface and groundwater of the area of Mali Gjere karst massif.

Sampling point	Name and type[a]	Altitude m	Mean discharge l/s	Number of samples	$\delta^{18}O$‰	δD‰	Conductivity μS/cm	Sulphate SO_4 mg/l
1 to 6	Blue Eye, s	152	18,400	34	-7.58 ± 0.16	-43.88 ± 2.4	585	135.0
7	Vrisi, s	177	70	8	-6.66 ± 0.13	-35.10 ± 2.0	430	32.2
9	Lefterohor, s	590	0.4	6	-6.95 ± 0.18	-37.00 ± 2.7	337	9.9
11	Kardhikaqi,s	185	90	8	-6.71 ± 0.16	-36.40 ± 1.6	567	121.4
12	Sopoti, s	1000	1.5	5	-8.26 ± 0.12	-47.20 ± 0.6	189	18.1
14	Viroi, s	196	0–35,000	5	-7.99 ± 0.19	-44.90 ± 2.9	371	47.0
15	Jorgucat, b	198	–	3	-6.80 ± 0.13	-42.20 ± 1.8	786	258
16	Drinos, r	197	–	3	-6.80 ± 0.13	-42.00 ± 1.7	950	385
17	Lista, s		1,700	–	–	–	300	–

[a]Type of sampling point: s, spring; b, borehole; r, river.

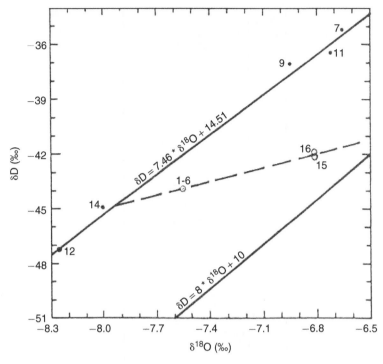

Figure 3. $\delta^{18}O$–δD plot of groundwater of Mali Gjere massif.

(no 14). Based on a simple two-component mixing analysis, the Drinos Valley gravely aquifer is estimated to contribute about 30% to the replenishment of the Blue Eye Spring. This is equal to $5.52\,m^3/s$ or $174*10^6\,m^3/year$.

2.4   *Discussion and interpretation of hydrochemical data*

The sulphate ion of the sampling points of the evaporation water line is used as a neutral ion for the estimation of the mixing proportions contributing to the Blue Eye Spring. The sulphate concentration varies about 300 to 500 mg/l in the Drinos River, about 250 to 300 mg/l in the Drinos Valley groundwater, while in the Blue Eye Spring it is about 135 mg/l.

The sulphate concentrations of two mixing components could be for the karst water 46.5 mg/l (Viroi Spring, no 14) and for gravely aquifer groundwater about 300. From two component mixing analysis we estimate that the contribution of the gravely aquifer to the recharge of the Blue Eye Spring is equal to about 35%, which is quite comparable with the value of 30% estimated with the isotopes.

The mean sulphate concentration of the Kardhikaqi Spring (no 11) of 121.4 mg/l is quite near to that of Blue Eye Spring. However, it seems that the origin of the sulphate ion of both springs is different. In fact, in the oxygen 18-deuterium plot (Figure 3) Kardhikaqi Spring is not on the mixing line. It seems that the gypsum deposits outcropping near this spring are responsible for its increased sulphate content.

2.5 *Mechanism of groundwater circulation in Mali Gjere karst*

During the rainy season (December–April), the karst groundwater level sensibly increases and a temporary water divide is created inside the karst massif. The karst groundwater flows at both directions, to the west and to the east (Figure 2). Beside of the large Viroi ephemeral spring (no 14), many other some hours to some days flowing duration springs appear at the eastern side of the Mali Gjere karst massif.

Usually from May month to October or November the karst groundwater level steadily decreases and the ground water flows mainly to the west of the massif, to the Blue Eye Spring. At that season all the springs of the eastern side of the massif dry up, including the Viroi Spring. The yearly amplitude of the karst water fluctuation observed in the Goranxi Cave is about 32 m. During the dry season the Drinos River is totally lost in the gravely aquifer which piezometric level suffers an unusual yearly decrease of about 20 to 25 m. Water level counters suggest the seepage of the gravely aquifer groundwater to the karst massif in the area Jorgucat – Dervican, and particularly around the Goranxi Cave (Figure 1).

3 PRESPA-OHRID LAKES AREA

3.1 *Geological and hydrogeological characteristics of study area*

The Small and Big Prespa and Ohrid Lakes share their water with Albania, Macedonia and Greece and constitute a common hydraulic system (Figure 4 and 5). The elevation of the Prespa Lake is 850 m a.s.l. and that of the Ohrid Lake is 695 m a.s.l., while the respective surfaces are 274 km^2 and 348 km^2. High mountains like the Mali Thate (2287 m) in the south and the Galichica (2,262 m) in the north separates them. Geologically these mountains represent a horst. On both sides of the horst are big graben structures; the Prespa Lake graben on the east, and the Ohrid Lake graben on the west. The mountain is constructed mainly of the Upper Triassic – Lower Jurassic massive limestone. Pliocene deposits of clay, sandstone and conglomerate fill most of the bottom of the Prespa and Ohrid lakes.

Because Mali Thate – Galichica Mountain is constructed of carbonate rocks, and the level of lakes has a difference of about 155 mm, a hypothesis was formulated (Cvijic 1906) that the Tushemisht and the St. Naun big karst springs issuing in the southern edge of Ohrid Lake are partially recharged by the Prespa Lake. The total discharge of the Tushemishit Spring is 2.5 m^3/s (79*10^6 m^3/year) and that of St. Naum Spring is 5.58 m^3/s (175*10^6 m^3/year). Some unknown water quantity drains in the Ohrid Lake. In the Zaver swallow hole located in Prespa Lake shore, the intensive loss of the lake water into the karst rocks could be observed.

The effective infiltration is accepted 55%, corresponding to 495 mm, the same as the value measured in the field in the Triassic limestone of neighbouring area of Kastoria in Greece (oral communication of A. Stamos, IGME – Kozani, Greece). The total karst groundwater resources estimated to be drain into Ohrid Lake is about 5.5 m^3/s or 173*10^6 m^3/year.

3.2 *Environmental tracer approach*

The altitude effect of the isotopic composition of the meteoric water is used for the identification of the waters coming from different potential groundwater recharge sources of the

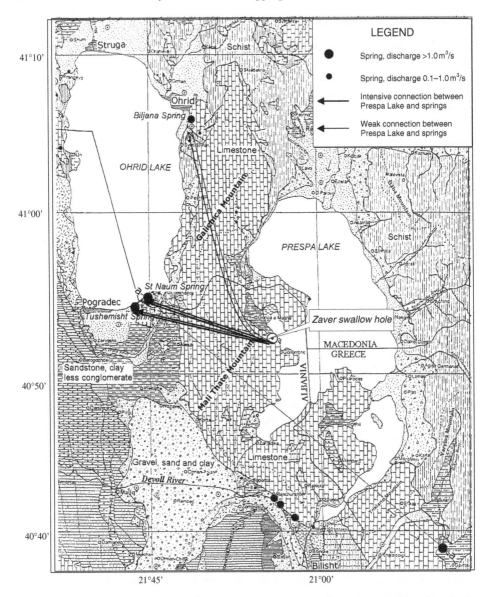

Figure 4.   Underground karst water connection between the Prespa Lake and the springs in the Ohrid lakeside.

study area. The local precipitation and Prespa Lake water are examined as some recharge potential source to Mali Thate – Galichica karst groundwater. The mean elevation of the groundwater recharge area is expected to be higher than the Prespa Lake elevation. The main analysed springs are shown in Figure 4. During the period 1989–1990, 5 to 12 samples are analysed for every sampling point in the Isotope Laboratory of IAEA in Vienna. The study is based on the results of isotope analyses of sampling points of the Albanian territory shown in Table 2 (Eftimi & Zoto 1997), but data from Macedonian territory (Anovski et al. 1991) are used also.

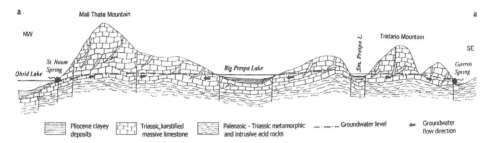

Figure 5. Cross-section a-a in Prespa-Ohrid karst massif (according Stamos, A. IGME; Kozani-Greece).

Table 2. Analyses for environmental isotopes in surface and groundwater of Prespa-Ohrid lakes area.

Sampling point	Name and type[a]	Altitude (m)	$\delta^{18}O$‰	δD‰	Deuterium excess (d)
A – Albanian territory (Eftimi & Zoto 1997)					
1	Ohrid Lake	695	−3.84 ± 0.43 (7)	−32.78 ± 2.70 (4)	−2.06
2	Prespa Lake	853	−1.72 ± 0.12 (9)	−21.84 ± 1.06 (9)	−8.08
3	Tushemisht – 1, s	696.5	−5.73 ± 0.0.07 (6)	−44.37 ± 0.68 (3)	+1.47
4	Zagorchan, s	698	−5.98 ± 0.07 (5)	−43.70 ± 1.77 (4)	+3.66
5	Tushemisht – 2, s	696	−5.80 ± 0.07 (6)	−43.93 ± 1.02 (3)	+2.47
6	Piskupat, s	700	−9.31 ± 0.16 (5)	−61.80 ± 0.60 (4)	+12.68
7	Big Golloborda, s	840	−9.76 ± 0.07 (7)	−63.885 ± 0.69 (4)	+14.23
8	Small Golloborda, s	880	−9.03 ± 0.04 (7)	−61.68 ± 1.73 (5)	+10.56
9	Manchurishta, s	848	−9.87 ± 0.07 (12)	−65.64 ± 1.09 (6)	+13.32
10	Progri, s	850	−9.20 ± 0.04 (11)	−62.14 ± 0.75 (5)	+11.46
11	Devoll, r	853	−8.87 ± 0.52 (5)	−59.65 ± 10.19(5)	+11.31
B – Macedonian territory (Anovski et al. 1991)					
1	St. Naum, p	695	−8.35	−52.9	+13.90
2	Stenje, p	853	−8.53	−55.4	+12.87
3	Velestovo, p	696.5	−8.74	−55.4	+14.82
4	Ohrid Lake	698	−3.4 ± 9.7	−35.4 ± 9.7	−8.2
5	Prespa Lake	696	−1.9 ± 0.30	−24.8 ± 9.6	−9.6
6	Galichica, s	700	−10.6 ± 0.6	−70.3 ± 5.4	+14.5
7	Biljana, s	840	−10.1 ± 0.5	−68.5 ± 4.0	+12.3
8	St Naum, s	880	−6.9 ± 0.3	−49.7 ± 3.0	+5.5

[a] Type of sampling point: s, spring; r, river; p, precipitation.

3.3 Discussion and interpretation of isotope data

The correlation function between mean $\delta^{18}O$‰ and δD‰ values of sampling points of Table 2 are presented in Figure 6. For characterising the Prespa and Ohrid lakes only the analytical data from the Albanian territory are used. The correlation function between mean $\delta^{18}O$‰ and δD‰ values of sampling points results in two equations, (2) and (3):

$$\delta D = 8\, \delta^{18}O + 14 \tag{4}$$

$$\delta D = 5.40\, \delta^{18}O - 12.42 \tag{5}$$

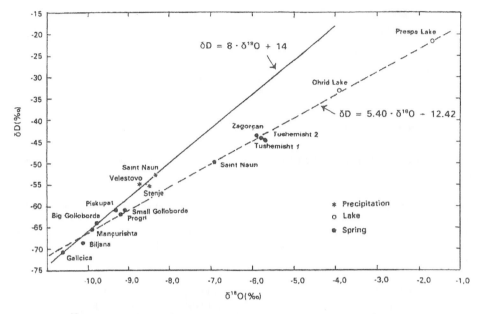

Figure 6. $\delta^{18}O–\delta D$ plot of groundwater and surface water of Prespa and Ohrid Lakes area.

The slope of the local meteoric water line is 8, the same as the world meteoric water line, but the intercept is 14 instead of 10. Bigger, "anomalous", values of intercept is a characteristic for Eastern Mediterranean countries (Gat & Dansgaard 1970, Gat & Galai 1982, Leontiadis et al. 1997, 1999).

A slope of about 4 to 6 is characteristic of water resources that have a significant high rate of evaporation relative to input. At the present case, the low slope of $\delta D-\delta^{18}O$ relationship of surface and groundwater is caused by the intensive evaporation of the Prespa Lake.

The mixing at different proportions of the infiltrated in the karst massif precipitation and of the Prespa Lake water is responsible for the isotopic composition of the springs falling in the mixing line. The mixing end-members are Prespa Lake (indexes $\delta^{18}O = -1.72‰$ and $\delta D = -21.84‰$), and the infiltrated in the karst massif precipitation represented by the point of the intercept of both lines of Figure 6 (indexes $\delta^{18}O = -10.20‰$ and $\delta D = -67.00‰$). From the two-component mixing analysis, we conclude that the Tushemisht Spring is recharged at 53% (1.3 m³/s) by the Prespa Lake and at 47% (1.2 m³/s) by the precipitations infiltrated in the Mali Thate-Galichca karst massif (Eftimi & Zoto 1997).

The contribution of Prespa Lake to the recharge of the St. Naun Spring is smaller; according to Anovski et al., (1991) it consists about 38% of the mean discharge of the spring. The total contribution of Prespa Lake to Tushemisht and St. Naun Springs is about 3.65 m³/s (115*10⁶ m³/year), and the contribution of the precipitation is about 4.45 m³/s (140.3*10⁶ m³/year).

3.4   *Artificial tracer experiment*

With the support of IAEA, an artificial tracer experiment was performed to further investigate the karst groundwater movement of Mali Thate-Galichica karst massif. Twenty kilograms of sulphorhodamine G Extra was injected on 18th of September 2002 at Zaveri swallow hole, in the Prespa Lake. The sampling campaign included some outlets of the St. Naun and Tushemisht Springs, as well as the Biljana Spring (Figure 4). The tracer concentration measured in Biljana Spring was incomparably smaller comparing with those measured in the other springs.

The maximum velocities vary from 233 m/h (Biljana Spring) to 3200 m/h (Tushemisht Spring). Slight differences of ground water flow velocity exist not only from one spring to another but even within the outlets of the same spring which testifies the presence of differently developed underground water passages at close distances. The maximal groundwater velocities measured in Prespa-Ohrid karst aquifer are comparable with the highest values of well-known karst areas of Croatia (Garašic 1997) and China (Kogovšek & Petric 1997). There is no doubt most of groundwater recharging the Tushemisht and St. Naun Springs circulates in well-developed big conduits.

4 GROUNDWATER VULNERABILITY AND PROTECTION

Mali Gjere and Prespa-Ohrid karst aquifers, like other similar aquifers are very vulnerable to pollution; however, both karst systems have also their particularities. In Mali Gjer karst aquifer, as Mandel (1967) describes in principal for the development of the karst processes, it seems that in the upspring direction from the Blue Eye Spring to the Goranxi Cave the huge diffuse flow system gradually is transformed into a conduit flow. The main pollution threat for the karst water of the Mali Gjere massif is potentially the Drinos River, which is vulnerable to main agriculture pollution in both Greek and Albanian territories. The monitoring and protection of karst water resources of Mali Gjere should be extended on entire Drinos River watershed and on Mali Gjere karst massif and a common programme might be applied on both bordering countries.

The main reason for concern in the Prespa-Ohrid karst aquifer is the concentrated point recharge in Zaver swallow hole followed by very rapid discharge to the St. Naun and Tushemisht Springs. The concentrated groundwater flow in big conduits limits the self-purification and the high flow velocities allow short transit times not enough for micro-organisms to die off (Drew 1999, Coxon 1999). The Prespa Lake is in the limit of eutrophication and the Ohrid Lake is susceptible to eutrophication (Watzin et al. 2002). The mean concentration of the total phosphorous in some monitoring stations of Prespa Lake varies from about 10 to 60 μg/l, which are considerably higher then those seen in the waters of Lake Ohrid. Unfortunately the determination of water quality (phosphorous and nitrogen in particular) in the big karst springs of St. Naun and Tushemisht recharging the Ohrid Lake has not yet received much attention.

5 CONCLUSIONS

The importance of environmental tracer methods (isotope and hydrochemical), as well as of artificial tracers, was revealed in the investigation of groundwater circulation in two transboundary carbonate aquifers in Albania. The karst groundwater circulation is very complicated. At both of the investigated aquifers we observed the intensive transfer of surface water and groundwater from one watershed to another. In the Prespa-Ohrid area the karst groundwater circulates mostly in well developed conduits, while in Mali Gjere both diffuse and conduit flow are important. The main pollution threat for the Blue Eay Spring of Mali Gjere massif is potentially the Drinos River, which is vulnerable to main agricultural pollution in both Greek and Albanian territory. The big karst springs of the Ohrid lakeside are vulnerable to pollution mainly from Prespa Lake water. The protection of both investigated aquifers requires the collaboration of the adjacent countries Albania, Greece and FYR of Macedonia.

REFERENCES

AHS (Albanian Hydrogeological Service) 1975. (Eds.) Eftimi R., Tafilaj I. & Bisha, G. Hydrogeological map of Albania scale 1:200 000. Tirana.

Anovski, T., Andonovski, B. & Minceva, B. 1991. Study of the hydrologic relationship between Ohrid and Prespa lakes, *Proceedings of IAEA International Symposium*, IAEA-SM-319/62p., Vienna, March, 1991.

Atkinson, T.C. 1977. Diffuse flow and conduit flow in limestone terrain in the Mendip Hills, Somerset (Great Britain). *Journal of Hydrology*, 35: 93–110.

Bradley, M., Brown, R.M., Gonfiantini, R., Payne, B.R., Przewlocki, K., Sauzay, G., Yen, C.K. & Yurtsever, Y. 1972. Nuclear techniques in groundwater hydrology. In: *Groundwater studies*. Chap. 10. UNESCO, Paris.

Civita, M. 1994. Vulnerability maps of aquifers subjected to pollution: theory and practice. Pitagora Editrice. Bologna (in Italian).

Coxon, C. 1999. Agriculturally induced impacts. In: *Karst hydrogeology and human activities*. A.A. Balkema, Rotterdam: 37–80.

Cvijic, J. 1906. Fundamental of Geography and Geology of Macedonia and Serbia, Special Edition VIII + 680, Belgrade (in Serb-Croat).

Daly, D., Dassargues, A., Drew, D., Dunne, S., Goldscheider, N., Neale, S., Popescu, I.C. & Zwahlen, F. 2002. Main concepts of the "European approach" to karst-groundwater-vulnerability assessment and mapping. *Hydrogeology Journal* 10: 340–345.

Drew, D. 1999. Introduction. In: *Karst hydrogeology and human activities*. A.A. Balkema, Rotterdam: 3–12.

Gat, J.R. & Dansgaard, W. 1970. Stable isotope survey of the freshwater occurrences in Israel and Jordan Rift Valley, *Journal Hydrology* 16; 177–212.

Gat, J.R. & Galai, A. 1982. Isotope Hydrology of Arava Valley: An isotope study of the origin and interrelationship Isr. J. Earth Sci., 31: 25–38.

Eftimi, R. & Zoto, J. 1997. Isotope study of the connection of Ohrid and Prespa lakes. In: *Proc. International Symposium "Towards Integrated Conservation and Sustainable Development of Transboundry Macro and Micro Prespa Lakes"*, Korcha, Albania, 24–26 October 1997: 32–37.

Garašic, M. 1997. Karst water tracing in some of the speleological features (caves and pits) in Dinaric karst area of Croatia: In: Kranjc, A. (ed.), *Tracer Hydrology*, Balkema-Rotterdam: 229–236.

Gogu, R.C. & Dassargues, A. 2000. Current trends and future challenges in groundwater vulnerability assessment using overlay and index methods. *Environmental Geology* 39(6): 549–559.

IAEA 1968. Stable isotope in hydrogeology, IAEA – Vienna, *Tech. Rep. Ser.* No 210.

Kessler, H. 1967. Water balance investigations in the karstic regions of Hungary. In: *Proc AIH-UNESCO, Symp. "Hydrolgy of Fractured rocks"*, Dubrovnik, 1965: 90–105.

Kogovšek, J. & Petric, M. 1997. Properties of underground water flow in karst area near Lunan in Yunan Province, China. In: In Kranjc, A. (ed.), *Tracer Hydrology*, Balkema-Rotterdam: 255–261.

Leontiadis, L.L.& Nikolau, E. 1999. Environmental isotopes in determining groundwater flow systems, northern part of Epirus, Greece. *Hydrogeology Journal* 7: 219–226.

Leontiadis, L.L, Smorniotis, C.H, Nikolau, E. & Georgiadis, P. 1997. Isotope hydrology study of the major areas of Paramythia and Korony, Epirus, Greece. In *Karst waters and environmental impacts*. A.A. Balkema, Rotterdam: 239–247.

Mandel, S. 1967. A conceptual model of karstic erosion by groundwater: In: *Proc AIH-UNESCO, Symp. "Hydrolgy of Fractured rocks"*, Dubrovnik, 1965: 662–664.

Motyka, J. 1998. A conceptual model of hydraulic networks in carbonate rocks, illustrated by examples from Poland. *Hydrogeology Journal* 6: 469–482.

Payne, B.R. Leontiadis, I., Dimitrulas, Ch. Dounas, A., Kallergis, G. & Morfis, A. 1978. A study of Kalamos Spring in Greece with environmental isotope. *Water Resources*, Vol. 14, No. 4: 653–658.

Shuster, E.T. & White, W.B. 1971. Seasonal fluctuation in the chemistry of limestone springs: a possible means for characterizing carbonate aquifers. *Journal of Hydrology* 14: 93–128.

Vrba, J., & Zaporozec, A. (eds.) 1994. Guidebook on mapping groundwater vulnerability. *International Contributions to Haydrogeology*, vol 16, Verlag Heinz Heise GmbH & Co KG, Hanover: p. 131.

Watzin, M.C. Puka, V. & Naumovski, T.B. 2002. Lake Ohrid and its watershed, State of the environment Report: p. 134.

White, W.B. 1969. Conceptual models for carbonate aquifers. *Ground Water* 7: 15–21.

CHAPTER 19

Karst aquifer intrinsic vulnerability mapping in the Orehek area (SW Slovenia) using the EPIK method

G. Kovačič[1] & M. Petrič[2]
[1] *University of Primorska, Faculty of Humanities Koper,Koper, Slovenia*
[2] *Karst Research Institute ZRC SAZU, Postojna, Slovenia*

ABSTRACT: The paper deals with the application of the EPIK parametric method to intrinsic vulnerability mapping in the Orehek area in southwestern Slovenia. In order to evaluate the option of reactivating the Korentan karst spring for usage as a reserve drinking water source, the intrinsic vulnerability map of its recharge area was produced. According to this method, values for four parameters (E, P, I and K) were assessed. By way of detailed litho-structural and geomorphological mapping in the field, special emphasis was placed on the assessment of parameters E and I. The final map shows that the vulnerability of the greater part of the aquifer is classified as "Moderate" to "High", while the vulnerability of only a small part of the aquifer is classified as "Very high". The influence of the parameters E and I on the calculation of the protection index F is significant. The importance of the aforementioned parameters was also proved through the sensitivity analysis, showing noticeable differences between theoretical and real weights.

1   INTRODUCTION

The Korentan karst spring near the town Postojna in southwestern Slovenia was included in the water supply system for this area up until 1972. Recently its reactivation as a reserve source has been discussed again due to some pollution problems with the sources which are used today. In order to evaluate this option, an intrinsic vulnerability map of the karst aquifer in the recharge area of the Korentan spring has been prepared as a basis for the determination of water protection zones.

2   HYDROGEOLOGICAL CHARACTERISTICS

The Korentan spring is the main outflow from the karst aquifer of the Orehek area built of Cretaceous and Palaeocene limestones (Figure 1). It is situated at the northern margin of this area at the contact with the surrounding relatively impermeable Eocene flysch. The limestone is highly karstified with typical surface and underground karst features. There are around 40 registered karst caves in this area and a significant number of dolines (which are mostly distributed along the main tectonic zones). On the basis of its hydrogeological

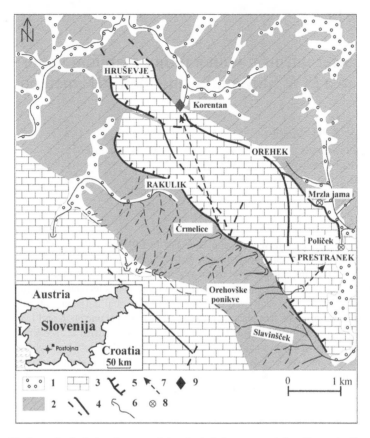

Figure 1.   Hydrogeological sketch map of the Orehek karst area (after Gospodarič et al. 1970) 1 – alluvial sediments, Holocene – intergranular aquifer, 2 – flysch rocks (marl and sandstone), Eocene – hydrogeological barrier, 3 – limestones and limestones with dolomites, Cretaceous and Palaeogene – karst aquifer, 4 – fault, 5 – thrust line, 6 – sinking stream, 7 – proved underground water connection, 8 – intermittent karst spring, 9 – discussed spring.

characteristics the carbonate unit is characterised as a highly permeable karst aquifer, and the surrounding flysch as a hydrogeological barrier. At the southern margin of the area, surface streams flowing over flysch sink into ponors at the contact with the limestone, and recharge the karst aquifer.

Two intermittent springs, the Mrzla jama and the Poliček, are situated at the eastern margin of the Orehek area. They are only active at very high groundwater levels, as at lower levels, the groundwater of this part of the area flows towards the river Pivka situated at a lower topographical level in the area. Such conditions have been confirmed by tracing tests. After injection of uranine into the Orehovške ponikve sinking stream, the tracer was detected in the Poliček spring and not in the Korentan spring (Gospodarič et al. 1970). During a second tracing test the Črmelice ponor was selected as the injection point, and groundwater flow towards the Korentan spring was proved, with an apparent velocity of 25 m/h (Schulte 1994). Based on these results, the calculation of hydrological balance and the geological structure of the area, the margins of the Korentan recharge area were defined. Its extent in the karst area has been estimated at $6\,\mathrm{km}^2$. In addition, around $0.2\,\mathrm{km}^2$ of the flysch area at the southern margin is included, where surface waters are collected.

The Korentan spring is a typical karst spring with rapid and sharp responses to precipitation events. High levels of precipitation of over $1,700\,\mathrm{mm}$ per year are characteristic of this area, and spring discharges range from several litres per second to more than $3\,\mathrm{m}^3/\mathrm{s}$. Water outflows through several fissures in the $50\,\mathrm{m}$ wide spring area in contact with the flysch. One of the fissures has been widened into a well, in which three pumps have been installed. During the active period from 1955 to 1972, on average around 15 to $20\,\mathrm{l/s}$ of water were pumped from this well into the water supply system. Chemical and bacteriological analyses of the spring have shown that the water complies with drinking water quality regulations.

3 FACTORS AFFECTING THE SELECTION OF THE EPIK METHOD

The concept of groundwater vulnerability assessment and mapping is founded on the assumption that the physical environment provides some mechanisms for the attenuation of potential natural and anthropogenic contamination. Since some areas are more vulnerable to groundwater contamination than others, the final goal of the concept is to produce a vulnerability map showing the catchment area of a particular drinking water resource subdivided into various more or less homogeneous areas describing the different levels of vulnerability (potential for contamination) (Vrba & Zaporozec 1994). Thus, various methods of mapping groundwater vulnerability have been developed and applied using different approaches and sets of parameters needed for vulnerability assessment. Parametric system methods represent the most common approach. The construction of the parameter system depends on the selection of those parameters considered to be representative as regards the assessment of groundwater vulnerability (Gogu & Dassargues 2000a).

Since karst regions are characterised as highly heterogeneous anisotropic environments with low self-cleaning capacity (natural remediation and neutralising potential), where watersheds are often very difficult to determine and variable in time, depending on the respective hydrologic conditions, intrinsic vulnerability mapping of karst aquifers is rather complex. Nevertheless, karst aquifer intrinsic vulnerability maps are widely used for the determination of water protection zones and, in combination with hazard maps, also as a practical instrument for land-use planning in highly vulnerable karst areas. In general, karst aquifer vulnerability mapping methods can be divided into those which are applicable to all types of aquifers, but providing additional methodological tools for karst, and those which are specially designed for use in karst regions, such as the EPIK method.

The EPIK method is a multi-attribute weighting-rating parametric system approach to intrinsic vulnerability mapping in karst terrains, using four parameters with assigned respective multipliers reflecting their relative importance in the final calculation of the protection index F: Epikarst (E), Protective cover (P), Infiltration conditions (I) and Karst network development (K) (Doerfliger et al. 1999). Since we wanted to map the areas with different degrees of vulnerability within the same karst aquifer, we decided to use the EPIK method. The other reasons for doing this were as follows: preliminary research showed that the weighting system is rather consistent with the field situation, giving greater importance to the I and E parameters; general absence of overlying layers (in particular subsoil and non-karst rocks); the method is appropriate for source vulnerability mapping, which is consistent with the concept of the protection of captured karst springs and wells in Slovenia; lack of sufficient data which determine the protective function of soil cover; the calculated protection index F can be easily transformed into protection zones with different water protection regimes.

4   USE OF THE EPIK METHOD IN THE OREHEK AREA

In the first step, the recharge area was divided into elemental cells, i.e. grid containing squares with 20 m long sides. Then each parameter was assigned a range of categories (three or four) and each cell was provided with the appropriate values related to each parameter. Epikarst characterisation (the parameter E) was based on topographic maps at 1:5,000 scale and detailed field mapping at the same scale. Geomorphological features such as dolines, caves, swallow holes, or karren fields were marked on topographic maps, and later their location and characteristics were verified in the field. Special emphasis was given to the detailed litho-structural mapping of the outcrops. According to the method of Čar (Čar 1982, Čar & Šebela 2001) and regarding the rock tectonic fracturing, three zones were defined: crush zones (characterised by tectonic clay, tectonic breccia and mylonite; virtually impermeable), broken zones (characterised by dense, chaotic systems of fault planes; moderately or highly permeable), and fissured zones (systems of fissures with negligible displacements; extremely permeable). All mapped features were drawn on topographic maps at 1:5,000 scale in a digital form. These maps were constructed in such a way that cells with abundant swallow holes, dolines, karren fields, fissured and broken zones were classified as the most vulnerable category E1 (Figure 2). Zones less dominated by these features were assigned category E2, and the rest of the catchment, the least vulnerable category E3.

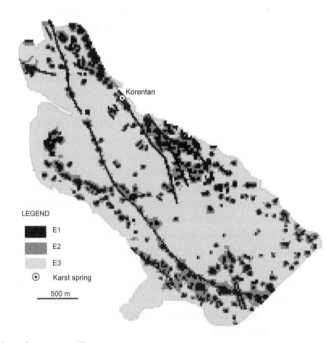

Figure 2.   Map of parameter E.

Basic data on the types and characteristics of soil cover were gathered from a 1:25,000 pedological map (Podatki tal Slovenije 2004) and pedological profiles (Lobnik 1991). For the entire studied area, the characteristics of the soils were rather uniform, so in estimating the parameter P, only the thickness of soil cover was considered. It was defined by field observation using the principle of morphological equivalence. Since no low hydraulic conductivity geological formation occurred below the soil, only three categories of parameter P (Figure 3) were assessed, based on soil thicknesses of 0–20 cm (P1), 20–100 cm (P2), and greater than 100 cm (P3).

Estimation of the parameter I was based on identification of zones of concentrated infiltration from topographic maps and field mapping. For areas with diffuse infiltration, the ground slope was calculated by the processing of DEM with a grid resolution of 25 m in the Golden Software Surfer 8. Types of land use were defined based on a vegetation map, a topographic map, and field observations. Four categories of parameter I were distinguished (Figure 4). Two cases were considered which correspond to the inner and outer areas of a stream catchment supplying a karstic swallow hole. Category I1 was assigned to cells with swallow holes, together with the banks and bed of streams recharging them. Inner areas with ground slopes greater than 10% for arable land and greater than 25% for meadows and pastures were characterised as I2, and those with lower slopes as I3. Outer areas with slopes greater than 10% and 25% respectively were also evaluated as I3, and the rest of the study area as I4.

To define parameter K, the presence of a karstic network in the aquifer and the degree of its development were assessed. Surface karst features distributed along fissured and broken fault zones indicate the presence of a well-developed karst system. In the catchment area of the Korentan spring around 40 caves are registered in the Slovene Cave

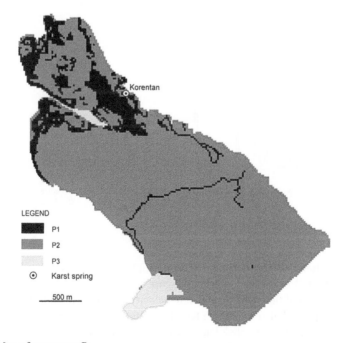

Figure 3.   Map of parameter P.

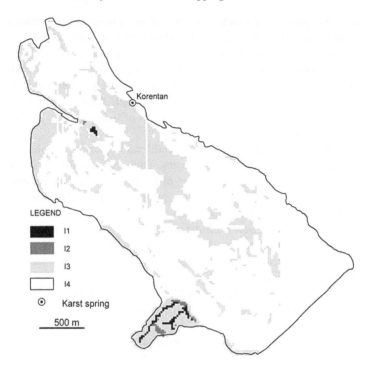

Figure 4.   Map of parameter I.

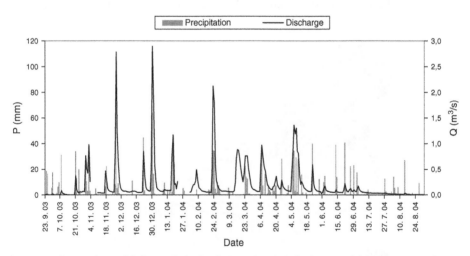

Figure 5.   Comparison of daily precipitation in Postojna and discharges of the Korentan spring.

Cadastre. In addition, indirect methods indicate a highly karstified aquifer. The reaction of discharges to rainfall events is very fast (Figure 5). A significant peak is followed by a rapid recession. Rapid flow with an apparent velocity of around 25 m/h at low to moderate water levels was proved also by tracing tests (Schulte 1994). These characteristics can be adopted for the whole of the Orehek karst area, so only category K1 for a moderate to well developed karstic network was defined.

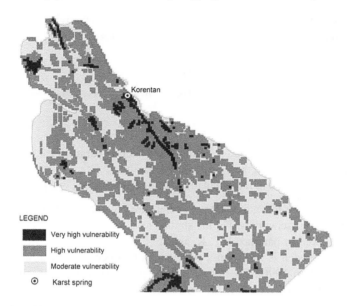

Figure 6.   Vulnerability map of the Korentan area.

The accuracy and reliability of the measured and assessed data required for vulnerability assessment are considered satisfactory. By using GIS tools and following the EPIK method the protection index F for each cell was calculated. Based on defined standards and in agreement with known characteristics of the studied aquifer, three vulnerability classes were defined: for F less than 19 very high vulnerability, for F between 20 and 25 high vulnerability, and for F greater than 25 moderate vulnerability. The final map (Figure 6) shows that the vulnerability of the greater part of the aquifer is classified as "Moderate" to "High", while the vulnerability of only a small part of the aquifer is classified as "Very high".

Comparing the maps for each parameter, the dominant influence of parameter E on the final map can be observed; however, the contribution of parameter I is also very important, especially for areas with concentrated infiltration. Due to the weighting system of the method, the influence of parameter P on the calculation of the protection index F is small.

5   SENSITIVITY ANALYSIS OF THE USED PARAMETERS

In order to analyse the influence of individual parameters and the results of the used EPIK method in the area of the Orehek karst, a sensitivity analysis was performed. To reduce the complexity of the analysis, the concept of "unique condition subareas" was applied (Napolitano & Fabbri 1996). Following this concept, all possible combinations of the parameters E, P, I and K were found and each such combination was defined as one subarea. In this way, 15.361 cells were reduced to 26 subareas. Amongst these 6 subareas covering less then 5 cells were not considered in further analyses.

During the first sensitivity test, the protection indexes F were calculated for each subarea from three parameters instead of four. To make the results comparable with the final protection index, the output values were multiplied by 4/3. Sensitivity for each parameter was calculated according to the following formula (Gogu & Dassargues 2000b):

$$SA = \left| \frac{F_i}{N} - \frac{F_{Ai}}{n} \right|_A$$

F_i – protection index in the subarea i
F_{Ai} – protection index in the subarea i without considering parameter A
N – number of maps used in the calculation (4 maps)
n – number of maps used in the calculation without considering one parameter (3 maps).

Statistic values of Sa for individual parameters are presented in Table 1. Higher values of Sa indicate greater sensitivity. For the studied area the sensitivity of the parameter I is the highest, followed by parameters E, K, and P.

The second approach is based on a comparison between theoretical and real weights (Napolitano & Fabbri 1996). The former are assigned using the EPIK method, and the latter are calculated according to the following equation:

$$W_{Ai} = \frac{A_{Ri} \cdot A_{wi}}{F_i} \cdot 100$$

A_{Ri} – rating of the parameter A assigned to the subarea i
A_{wi} – weight of the parameter A assigned to the subarea i
F_i – protection index in the subarea i.

The results of the comparison are shown in Table 2. Noticeable differences between theoretical and real weights can be seen. The value of the real weights is significantly higher

Table 1.   Statistic values of sensitivity Sa.

Parameter	Average	Standard deviation (%)	Median (%)	Minimum (%)	Maximum (%)
SE	1.12	0.69	1.13	0.08	2.33
SP	0.99	0.49	0.92	0.08	1.92
SI	1.27	0.71	1.21	0.08	2.50
SK	1.06	0.41	1.04	0.25	1.75

Table 2.   Statistical analysis of the weight parameter sensitivity.

Parameter	Theoretical weight	Theoretical weight (%)	Average weight (%)	Standard deviation (%)	Median (%)	Minimum (%)	Maximum (%)
E	3	33.33	36.38	13.94	40.02	15.00	60.00
P	1	11.11	11.56	5.68	11.54	3.70	27.27
I	3	33.33	41.79	13.81	42.86	15.00	66.67
K	2	22.22	10.26	2.81	9.76	6.90	18.18

for parameter I, and slightly higher for parameter E. The difference between the values is also significant for parameter K – the reason for this can be accounted for by the fact that a constant value of the K parameter was applied in the calculation of the vulnerability index.

Based on the comparison between the maps obtained for each individual parameter and those relating to real weights, a significant influence of parameters I and E on the real weights can be highlighted. The high real weights of parameter I are related to highly vulnerable areas characterised by low values of parameter E. The high real weights of parameter E are typical of all areas with categories of parameter I other than I4. The real weights of parameter P are particularly influenced by the values of parameter I. It can be seen that the real weights are conditioned by the values of a single parameter as well as by other parameters.

It can be concluded that in the Orehek area, estimation of vulnerability using the EPIK method was most sensitive to the parameters I and E. The results obtained emphasise the importance of a careful and exact assessment of these two parameters.

6 CONCLUSIONS

By utilisation of the EPIK method, an intrinsic vulnerability map of the Orehek area has been prepared. As also suggested by a sensitivity analysis of the parameters, special emphasis was given to the assessment of the parameters E and I. Existing data and carto-graphic material were complemented with detailed litho-structural and geomorphological mapping in the field. In this way, the quality of the input information was improved, and consequently the quality of the intrinsic vulnerability map was considered to be enhanced.

To test and confirm such improvement, further research has been planned. Data-loggers for measuring discharge and specific electric conductivity, temperature and pH of water have been installed at the Korentan spring in order to learn more about the hydrodynamic functioning of the karst aquifer. The next step will be to conduct combined tracing tests in the Orehek area. The comparison and processing of gathered data will form the basis for a validation of the utilised method for intrinsic vulnerability mapping.

REFERENCES

Čar, J. 1982. Geološka zgradba požiralnega obrobja Planinskega polja (Geologic Setting of the Planina Polje Ponor Area). *Acta carsologica* 10 (1981): 75–105.
Čar, J. & Šebela, S. 2001. Kraške značilnosti narivnega stika apnenec-dolomit pri Predjami (Karst characteristics of thrust contact limestone-dolomite near Predjama). *Acta carsologica*, 30/2: 141–156.
Doerfliger, N., Yeannin, P.-Y. & Zwahlen, F. 1999. Water vulnerability assessment in karst environ-ments: a new method of defining protection areas using a multi-attribute approach and GIS tools (EPIK method). *Environmental Geology* 39 (2): 165–176.
Gogu, R.C. & Dassargues, A. 2000a. Current trends and future challenges in groundwater assess-ment using overlay and index method. *Environmental Geology*, 39 (6): 549–559.
Gogu, R.C. & Dassargues, A. 2000b. Sensitivity analysis for the EPIK vulnerability assessment in a small karstic aquifer, southern Belgium. *Hydrogeology Journal* 8 (3): 337–345.
Gospodarič, R., Habe, F. & Habič, P. 1970. Orehovški kras in izvir Korentana (The karst of Orehek and the source of the Korentan). *Acta carsologica*, 5: 95–108.
Lobnik, F, 1991. Pedološka karta Republike Slovenije, sekcija Postojna – opis pedoloških profilov. Printed research report, NUK Ljubljana.

Napolitano, P. & Fabbri, A.G. 1996. Single-parameter sensitivity analysis for aquifer vulnerability assessment using DRASTIC and SINTACS. HydroGIS '96: Application of GIS in Hydrology and Water Resources Management *Proceedings of the Vienna Conference, IAH Publ*, no. 235: 559–566.

Podatki tal Slovenije 2004. Center za pedologijo in varstvo okolja. URL (quoted 15.1.2004): http://www.bf.uni-lj.si/cpvo/Novo/main_PodatkiTalSlovenije.htm.

Schulte, U. 1994. Geologische und Hydrogeologische Untersuchungen im Karst von Orehek (Slowenien). Diplomarbeit, Universität Karlsruhe, Deutschland.

Vrba, J. & Zaporozec, A. 1994. Guidebook on mapping groundwater vulnerability. International association of hydrogeologists. Hannover.,Verlag Heinz Heise.vol. 16: p. 31.

CHAPTER 20

Physically-based intrinsic groundwater resource vulnerability map of the Tisovec karst

P. Malík[1] & S. Vojtková[2]

[1] Geological Survey of Slovak Republic, Bratislava, Slovak Republic
[2] Comenius University, Faculty of Natural Sciences, Department of Hydrogeology Bratislava, Slovak Republic

ABSTRACT: Within this paper, an application of groundwater vulnerability assessment on the hydro-geological structure of the Tisovec karst Slovenske Rudohorie Mts. (Western Carpathians – Slovakia) is described. For calculation of physically-based vulnerability parameters, the VULK tool developed and applied at University of Neuchâtel in Switzerland was used. The target of a possibly spreading contamination was the groundwater table, so the final map is an intrinsic groundwater resource vulnerability map due to the COST 620 project definitions. The 20.45 km^2 of the outcropping limestones and dolomites were cover by a 100×100 m grid, counting 12,644 cells. Values of 4 basic parameters (soil thickness, soil hydraulic conductivity, unsaturated zone (epikarst) thickness, unsaturated zone (epikarst) hydraulic conductivity) were derived from soil samplings, geological and hydrogeological maps and digital elevation model (DEM). All the data were classified into several coded categories, counting 62 combinations from 72 theoretically possible on the territory of Tisovec karst. Runs of VULK tool gave simulated contamination breakthrough curve parameters such as t_{break}, $t_{duration}$ and C_{max}/C_0 values. These were logarithmically plotted on the X_1, X_2 and X_3 axes of the "vulnerability cube". In the case of the Tisovec karst the "zero point", i.e. the most vulnerable point of the vulnerability cube was given as $C_{max}/C_0 = 1$; $t_{break} = 1 \times 10^{-2}$ day; $t_{duration} = 1 \times 10^5$ days. The final vulnerability value V is calculated as the distance from the "zero point" of the "vulnerability cube". Values of t_{break}, $t_{duration}$ and C_{max}/C_0 as well as of "vulnerability value" V were connected to the initial grid to create intrinsic groundwater resource vulnerability map of the Tisovec karst.

1 INTRODUCTION

In the framework of Action COST 620 project (Zwahlen ed. 2004), new "European approach" in groundwater vulnerability assessment was introduced by hydrogeological teams from 15 European countries. Within the COST 620 action, new definitions of groundwater vulnerability (source, resource, intrinsic, specific …) as well as new physically-based groundwater vulnerability concept. For intrinsic vulnerability, three factors describing pollution by a conservative contaminant were defined: contaminant transfer time (t_{break}), duration of contamination ($t_{duration}$) and level of concentration reached by the contaminant (C_{max}/C_0) – parameters of a breakthrough curve. Nevertheless, the evaluation of groundwater vulnerability using physically-based approaches (Brouyére et al. 2001, Jeannin et al. 2001) does not have many applications yet.

Within this paper, an application of groundwater vulnerability assessment on the hydrogeological structure of the Tisovec karst is described. For calculation of physically-based vulnerability parameters, the VULK tool developed and applied at University of Neuchâtel in Switzerland (Cornaton & Perrochet 2001) was used. The target of a possibly spreading contamination was the groundwater table, so the final map is an intrinsic groundwater resource vulnerability map due to the COST 620 project definitions.

2   TISOVEC KARST IN SLOVAKIA – BRIEF DESCRIPTION

Tisovec karst is a part of a Mesozoic carbonate rim of a large crystalline and Paleozoic mass of the Slovenské Rudohorie (Slovak Ore Mts.) in the central part of Slovak Republic (Figure 1). Built by Middle and Upper Triassic limestones and dolomites, it creates karstic area with the extent of 20.45 km² N and W from the city of Tisovec, between two surface streams of Furmanec a Rimava, or W from them, respectively. The geological structure of the Tisovec karst is quite complex, formed by several overthrusted Mesozoic units (Foederata series, Turnaicum and Silicikum Units), later complicated by younger vertical and horizontal tectonic movements. Based on geological structures, Tisovec karst was subdivided into two hydrogeological structures Tisovec karst s.s. and Kučelach massive; (Wiesengangerová, 2000), which, according to Kullman's classification (1990), can be classified as closed hydrogeological structures. Within this paper, only Tisovec karst s.s. hydrogeological structure was evaluated. In this structure, Wetterstein dolomites, Wetterstein limestones and Steinalm limestones Middle Triassic lithological facies can be found, as well as Upper Triassic Dachstein limestones and Tisovec limestones of the Silicicum Unit, represented by the nappe of Murán. Outcrops of Upper Triassic carbonates prevail in this structure, which makes it different from the majority of Slovak karst aquifers. Foederata series and Turnaicum units are overlaid by Silicium and their outcrops are relatively small.

The most important karstic springs of the Tisovec karst s.s. hydrogeological structure are the "Teplica" spring with average discharge of 73.5 l/s (gauging in the period of 04.01.1956–13.9.1967; max. ca 700 l/s; min. 5.25 l/s) and "Periodic spring", where average discharge in the period of 09.04.2001–15.04.2001 and 25.06.2001 – 01.07.2001 was 25.39 l/s, minimal discharge is from 6.0 to 8.0 l/s; and maximal from 45 to 50 l/s (Padúch 1998). "Periodic spring" is one of the two intermittent karstic springs, registered on the

Figure 1.   Position of Tisovec karst on the Slovak territory.

Slovak territory. This spring has three outlets – one permanent and two intermittent. Groundwater leaves permanent outlet continuously, even in the dry periods, when the discharge falls to ca $61 \times s^{-1}$. Reaching maximal discharge impulse from the minimum last approximately 6 minutes, maximal state another 3 minutes and the gradual decrease towards minimum 37 minutes. The first (I.) and second (II.) intermittent outlet are active only during the maximal discharging period. Outlet II. starts 2 min. after the outlet I. (Michalko & Vojtková 2003). Cave with three interconnected underground lakes was found by speleological survey in the limestone slope behind the "Periodic spring" (Wiesengangerová 2000). Both "Teplica" and "Periodic spring" are used for water supply of the city of Tisovec. "Periodic spring" is used since 1959 and serves as the main source. "Teplica" spring acts as a supplementary source only during the dry periods, when "Periodic spring" average discharge decreases.

3 DATA PROCESSING FOR GROUNDWATER VULNERABILITY ASSESSMENT

The 20.45 km^2 of the outcropping limestones and dolomites were cover by a $100 \times 100 \text{ m}$ grid, counting 12,644 cells. Values of 4 basic parameters (soil thickness, soil hydraulic conductivity, unsaturated zone (epikarst) thickness, unsaturated zone (epikarst) hydraulic conductivity) were derived from soil samplings, geological and hydrogeological maps and digital elevation model (DEM). After the soil sampling it was clear that the soil thickness depends strongly on the terrain slope, and DEM parameters were used to delineate typical slope values. Two soil types were identified according to granulometric analyses, but the values of soil hydraulic conductivity were derived not from these analyses, as these values were found unreliable by Adamcová et al. (2002, 2003). Instead, analogical values measured in the field in similar soil types by these authors, were exploited. Values of unsaturated zone thickness (epikarst zone in the karstified limestone areas) were derived by DEM as well, using the major spring of the "Periodic spring" as the altitudinal base for groundwater table, similarly as it was applied in L'Abbaye case (Anonymous, 2001). Hydraulic conductivity values for rocks were derived from hydrogeological map. For simplification of the number of VULK tool simulations, all the data were classified into coded categories. Parameter combinations were coded using 4 alphanumeric digits (□□□□). The first code digit is for soil thickness interval, the second digit for soil hydraulic conductivity interval, the third code number is for unsaturated zone thickness interval and the last, fourth digit, for unsaturated zone hydraulic conductivity interval (Table 1). Resulting combination of categories occurring on the territory of Tisovec karst gave 62 combinations from 72 theoretically possible.

Breakthrough curve simulations using VULK tool require flow velocity values, which were calculated by using hydraulic conductivity values and unit hydraulic gradient value (more relevant method should be implemented later). Single porosity of 0.1 (10%) was used; secondary porosity was neglected. Dispersivity value vas calculated as 1/20 of the flow distance (layer thickness). Because contaminant arrival was "traced" on the groundwater table (resource vulnerability principle according to COST 620 classification, (Zwahlen ed. 2004), zero dilution entered the simulation process as well. For all simulated contaminant breakthrough cases (parameter code combinations listed in Table 1), input time interval was set as 1 hour.

Breakthrough curves of a conservative contaminant were simulated using the VULK-tool (Cornaton & Perrochet 2001; Jeannin et al. 2001), with the input parameters representing

Table 1.   Input parameters of individual subsystems for breakthrough curve modelling.

Code number	Description	Real parameter values
	X□□□ – soil thickness [m]	
0	Steep slope	0.1 m
1	Moderate slope	0.5 m
2	Plain	1.0 m
	□X□□ – soil hydraulic conductivity [m/s]	
0	Kambisoils	1E-05 m/s
1	Kambisoils	1E-05 m/s
2	Luvisoils	3E-06 m/s
	□□X□ – epikarst thickness [m]	
0		50 m
1		50 m
2		150 m
3		250 m
4		350 m
5		450 m
6		550 m
7		650 m
	□□□X – epikarst hydraulic conductivity [m/s]	
0	Non-carbonate rocks	3E-05 m/s
1	Limestones	3E-04 m/s
2	Dolomites	1E-04 m/s

each of these categories gave adequate values (Table 2). Shapes of typical simulated break-through curves are shown on Figure 2. Based on six breakthrough curves for 6 parameters combination from Figure 2, the time development of contaminant relative concentration on groundwater table level in hours after contamination release on surface is shown on Figure 3.

The size of circles there represents values of relative concentration; time is shown on the vertical axes. It is clear from Figure 3, that the first two cases (relatively thin unsaturated zone of limestones) can create high concentrations, but quickly diminishing, while the cases with dolomites (less permeable rocks) or with thick unsaturated zone in limestones have the same effect on concentration. Its duration is more clearly shown on Figure 2.

Resulting t_{break}, $t_{duration}$ and C_{max}/C_0 values were logarithmically plotted on the X_1, X_2 and X_3 axes of the "vulnerability cube" defined by Brouyére et al. (2001) (Figure 4). Final vulnerability value V is then calculated as a position – distance – of a point determined by its X_1, X_2 a X_3 coordinates (or C_{max}/C_0, t_{break} and $t_{duration}$ values) from the "zero point" of a "vulnerability cube" according to equations (1, 2, 3 and 4).

$$V = (X_1^2 + X_2^2 + X_3^2)^{1/2} \tag{1}$$

$$X_1 = -\log(C_{max}/C_0) \tag{2}$$

$$X_2 = 5 - \log(t_{duration}) \tag{3}$$

$$X_3 = 2 + \log(t_{break}) \tag{4}$$

where:

V – final groundwater vulnerability value of the given point of the "vulnerability cube"

X_1 – position on the "relative concentration" axes of the "vulnerability cube"

Table 2. Simulated values of breakthrough time t_{break}, duration time $t_{duration}$ and maximum relative concentration C_{max}/C_0 of a contamination for individual combination of parameters.

input parameter's code	1100	1101	1102	1110	1111	1112	1120	1121	1122	1130	1131	1132	1140	1141	1142	1150
t_{break} [h]	46.95	6.13	15.38	46.95	6.13	15.3	137.	15.3	43.1	228	24.6	70.9	319	33.9	98.7	410
$t_{duration}$ [h]	38.32	7.69	17.48	38.32	7.69	17.4	19	17.4	36.5	0	25.2	45.7	0	31.45	45.97	0
C_{max}/C_0	0.031	0.273	0.1	0.03	0.27	0.1	0.01	0.1	0.03	0.006	0.060	0.020	0.004	0.043	0.014	0.003

input parameter's code	1151	1152	1160	1161	1170	1171	2200	2201	2202	2210	2211	2212	2220	2221	2222	2230
t_{break} [h]	43.16	126.5	501.5	52.42	529.4	61.68	55.95	15.13	24.38	55.95	15.13	24.38	146.9	24.38	52.16	237.8
$t_{duration}$ [h]	36.55	33	0	40.59	0	43.64	39.01	14.96	20.94	39.01	14.96	20.94	17.75	20.94	37.37	0
C_{max}/C_0	0.034	0.011	0.003	0.028	0.002	0.023	0.03	0.123	0.078	0.003	0.123	0.078	0.010	0.078	0.032	0.006

input parameter's code	2231	2240	2241	2250	2251	2252	2260	2261	2270	0000	0001	0002	0010	0011	0012	0020
t_{break} [h]	33.65	328.7	42.91	419.6	51.66	135.5	510.5	61.42	601.4	46.15	5.33	14.59	46.15	5.33	14.58	137.0
$t_{duration}$ [h]	27.21	0	32.72	0	37.37	32.92	0	41.11	0	38.32	7.71	17.41	38.32	7.71	17.41	19
C_{max}/C_0	0.054	0.004	0.041	0.003	0.032	0.011	0.003	0.027	0.002	0.031	0.29	0.10	0.031	0.29	0.10	0.010

input parameter's code	0021	0022	0030	0031	0032	0040	0041	0042	0050	0051	0052	0060	0061	0071
t_{break} [h]	14.58	42.36	228.0	23.85	70.14	318.8	33.12	97.92	409.8	42.36	125.7	500.7	51.62	60.88
$t_{duration}$ [h]	17.41	36.54	0	25.15	45.74	0	31.44	45.97	0	36.54	33.45	0	40.58	43.64
C_{max}/C_0	0.101	0.034	0.006	0.061	0.020	0.004	0.043	0.014	0.003	0.033	0.011	0.003	0.027	0.023

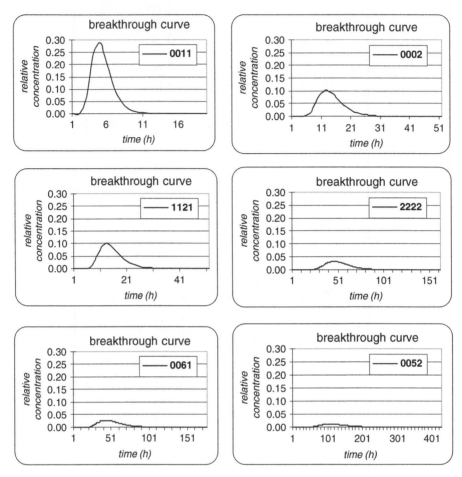

Figure 2.   Shapes of typical breakthrough curves simulated by VULK for the limestones (0011 – thickness of the soil cover bs = 0.1 m, hydraulic conductivity of the soil cover ks = 1E-05 m/s, thickness of the unsaturated (epikarst) zone be = 50 m, hydraulic conductivity of the unsaturated (epikarst) zone ke = 3E-04 m/s; 1121 – bs = 0.5 m, ks = 1E-05 m/s be = 150 m, ke = 3E-04 m/s; 0061 – bs = 0.1 m, ks = 1E-05 m/s, be = 550 m, ke = 3E-04 m/s) and dolomites (0002 – bs = 0.1 m, ks = 1E-05 m/s, be = 50 m, ke = 1E-04 m/s; 2222 – bs = 1 m, ks = 3E-06 m/s, be = 150 m, ke= 1E-04 m/s; 0052 – bs = 0.1 m, ks = 1E-05 m/s, be = 450 m, ke = 1E-04 m/s).

X_2 – position on the "mean contamination duration time" axes
X_3 – position on the "mean contaminant transfer time" axes of the "vulnerability cube"
C_0 – relative input concentration on the surface [−]
C_{max} – maximal contaminant concentration in the evaluated point [−]
$t_{duration}$ – mean duration time of contaminant relative concentration over threshold value [D]
t_{break} – mean contaminant transfer time [D]

It is clear, that the final vulnerability value depends on the "zero point", or the point of maximal vulnerability within the "vulnerability cube". In the case of the Tisovec karst

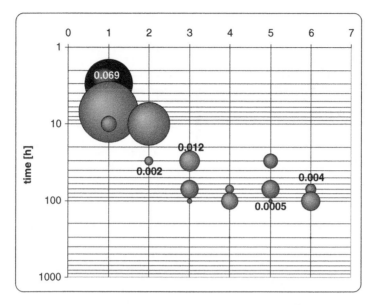

Figure 3. Development of contaminant relative concentration on groundwater table level in hours after contamination release on surface (based on breakthrough curves depicted on Figure 2).

The size of circles represents relative concentration values, time is on the vertical axes and six marked cases shown on horizontal axes are as follows: 1 − [0011] 0.1 m of kambisoils + 50 m of limestones; 2 − [1121] 0.5 m of kambisoils + 150 m of limestones; 3 − [0061] 0.1 m of kambisoils + 550 m of limestones; 4 − [0020] 0.1 m of kambisoils + 50 m of dolomites; 5 − [2222] 1.0 m of luvisoils + 150 m of dolomites; 6 − [0052] 0.1 m of kambisoils + 450 m of dolomites.

(on the contrary with the "vulnerability cube" zero point given by Jeannin et al. 2001), as can be seen also from the equations (1; 2 and 3), the "zero", i.e. the most vulnerable point of the vulnerability cube was given as $\mathbf{C_{max}/C_0} = 1$; $\mathbf{t_{break}} = 1.10^{-2}$ day; $\mathbf{t_{duration}} = 1.10^5$ days. The final vulnerability value $\mathbf{V}$ is calculated as the distance from the "zero point" of the "vulnerability cube".

4 "VULNERABILITY CUBE" FOR THE TISOVEC KARST

Contaminant relative concentration values $\mathbf{C_{max}/C_0}$ simulated for 62 combinations of the input parameters in the Tisovec karst (Table 2; Figures 2, 3 and 4) are within the range of 0.002 to 0.290 with the mean value of 0.040. Plotted on the X_1 axes of the "vulnerability cube" using the equation (2), they are within the interval of 0.538 to 2.622 with the mean value of 1.721. As a threshold value for the relative concentration of contaminant the value of 0.01 (or 1%) was considered, taking into account typical situations and classical conservative contaminant (e.g. nitrates) scenario. Values of the mean contamination duration time $t_{duration}$ with the relative concentration > 0.01 for 62 breakthrough curves, calculated by VULK-tool were within the range from 0.000 (no contamination) up to 45.973 hours (1.92 days), with the average value of 22.641 hours (0.94 days). Position of the mean contamination duration time on the X_2 axes of the "vulnerability cube" was within the interval of

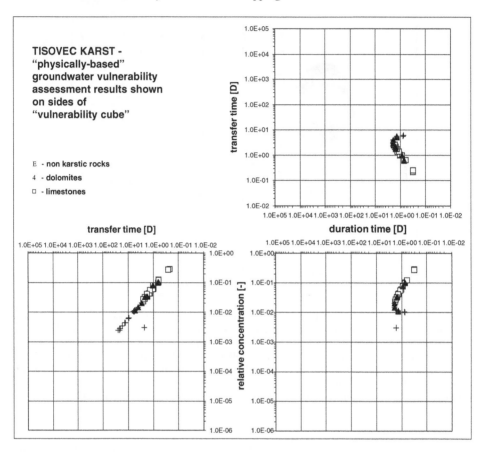

Figure 4.   Plotting of simulated values of breakthrough time t_{break}, duration time $t_{duration}$ and maximum relative concentration C_{max}/C_0 of a contamination for individual combination of parameters on the "vulnerability cube" for the Tisovec karst.

4.718–7.000 (highest values were given to the zero contamination) with the average value of 5.476 (Table 2; Figures 2 and 4). Values of the mean contamination transit time t_{break} for the case of the Tisovec karst were from 5.33 to 601.41 hours (0.22–25.0 days) with the average of 142.54 hours (5.94 days). Position of the mean contamination transit time on the X_3 axes of the "vulnerability cube" was within the interval from 1.346 to 3.399 with the average of 2.498. Final groundwater vulnerability value for the Tisovec karst **V** calculated according to eq. (1) as a distance of the point determined by C_{max}/C_0, t_{break} and $t_{duration}$ from the "zero point" according to eq. (1, 2, 3 and 4) was ranging from 5.501 to 8.211 with the mean value of 6.281 (Figure 4).

Based on physical values of theoretical contaminant spreading in the rock environment of the Tisovec karst, simulated as breakthrough curves by the VULK-tool, 3 source maps were compiled for the evaluated area: (a) map of maximal relative concentrations of the conservative contaminant, (b) map of contamination mean breakthrough time and (c) map of mean contamination duration time over accepted threshold limit. Final groundwater vulnerability map for the Tisovec karst was then created by linking of **V** values

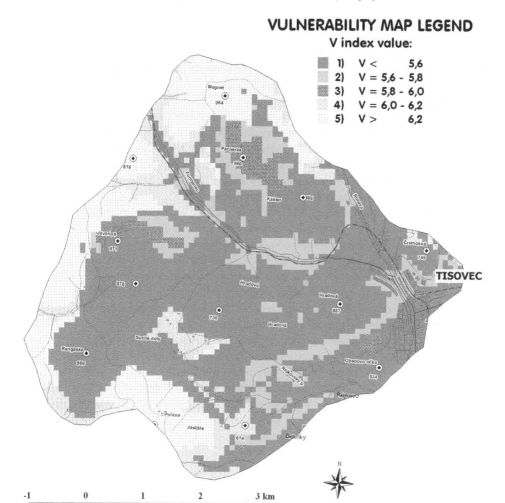

Figure 5.   Resulting intrinsic resource groundwater vulnerability map of Tisovec karst hydrogeological structure.
1 – very high vulnerability, 2 – high vulnerability, 3 – moderate vulnerability, 4 – low vulnerability, 5 – very low vulnerability.

$(V = (X_1^2 + X_2^2 + X_3^2)^{1/2})$ to the initial cells of the grid and by further re-classification of these values to five categories. Final classification of groundwater vulnerability of the Tisovec karst was within the interval limits of $- 5.6 - 5.8 - 6.0 - 6.2 -$ with names devoted to these categories: very low $-$ low $-$ moderate $-$ high $-$ very high vulnerability (Figure 5). The final intrinsic groundwater resource vulnerability map of the Tisovec karst was plotted in the 1:50,000 scale afterwards.

It is clear from the simplification of wide range of input data at the beginning (Table 1), that the final product, the intrinsic groundwater resource vulnerability map can characterise only limited number of simulated cases of a schematised environment. However, it brings into light the real possibilities of pollution spreading within the Tisovec karst. The final intrinsic groundwater resource vulnerability map (Figure 5) shows, that groundwater is most vulnerable in the areas built by limestones in lower altitudes (areas of Hradová, Káster and Rangaska), slightly less vulnerable is groundwater along the Furmanec stream in the area of the "Periodic spring" and Teplica spring, in the Rejkovsky potok valley and in the NE part of the Tisovec surroundings. NE slopes of the Pacherka hill show moderate vulnerability values. The best natural groundwater protection (lowest vulnerability) is in the crystalline rim of the Tisovec karst and in the Lower Triassic rocks of the Suché doly area.

When comparing different rock environments of limestones and dolomites and simulated breakthrough curve results, respectively, values of maximal relative concentrations C_{max}/C_0 for limestones are ranging from 0.23% to 29.0%, and for dolomites from 1.10% to 10.1% with mean values of 9.5% for limestones and 4.8% for dolomites. Mean contaminant transit time t_{break} for limestones was 0.22–2.57 days with 1.31 days average, for dolomites 0.61–5.65 days with the average of 2.53 days. Contamination duration time $t_{duration}$ over accepted threshold limit (>1%) was for limestones 0.32–1.82 days (average 1.11 days) and for dolomites 0.73–1.92 days (average 1.32 days). Such results support empirical knowledge on fast and concentrated contaminant arrival with shorter duration in limestones and less concentrated later contaminant arrival with longer duration in dolomitic rock environment (Table 2, Figure 4). Large range of values (Figure 4) within equal rock environments points on dominant influence of the unsaturated zone thickness on the groundwater vulnerability.

The applied concept of groundwater vulnerability assessment is based on physically based model of contamination spreading. However, this model was utilised only in the most simplified way, in order to cope with the number of model simulations, which had to be performed manually for 62 schematised cases – combinations of 4 parameters' ranges. In the future, it should be easier to run an automatic simulation process for the real parameters present in each grid cell. As for now, simulation is run only for the "worst case", i.e. fully saturated soil (e.g. after a heavy rain or snow melt), and only percolating part of water, excluding surface runoff is taken into account. In the karstic environment, it often happens that a part of surface runoff flows into swallow holes and sinkholes, thus entering the underground water system in a concentrated manner. Such cases can be described by breakthrough curves as well, based again on physical principles of contamination spreading. Such phenomena were not described in this paper, but the used approach is a very promising in building up a comparable (using defined physically defined parameters) vulnerability mapping methodology. As the use of several different "point-counting" methods can bring contradictory results for thesame area, this physically based method is worth of further development. The resulting groundwater vulnerability values should then depend only of the precision of knowledge of respective parameters in the field, and can be improved only by considering more precise data and data processing. The validity of the final groundwater vulnerability map can be checked by "simple" comparisons of modelled processes and the measured reality on available sites. Both results of tracing test, but also results of all experiments from which the unsaturated zone / aquifer properties can be obtained, can serve as a validation tool.

REFERENCES

Anonymous 2001. Excursion guide, Theme 2.3, Cartography of the Vulnerability, European Approach and VULK application, L'Abbaye. 7th conference of Limestone Hydrology and

Fissured Media. Besacon (France), University of Franche-Comté and University of Neuchâtel, 20th–22nd September 2001.

Adamcová, R., Durn, G., Miko, S., Skalský, R., Kapelj, S., Dubíková, M. & Ottner, F. 2002. Soil cover as pollution barrier in karst areas – brief summary of research results. Jakosc i podatnosc wod podziemnych na zanieczyszczenie. *Prace Wydzialu Nauk o Ziemi Uniwersytetu Slaskiego* Nr. 22, Sosnowiec 2002: 9–17.

Adamcová, R., Ottner, F., Durn, G., Greifeneder, S., Dananaj, I., Dubíková, M., Skalský, R., Kapelj, S. & Miko, S. 2003. Hydraulic conductivity of karst soils. In: *Proc. of Conference on Applied Environmental Geology (AEG'03) Wien*, Umweltbundesamt/Federal Environment Agency – Austria, BE 228 (2003): 6–8.

Brouyére, S., Jeannin, P.-Y., Dassargues, A., Goldscheider, N., Popescu, C., Sauter, M., Vadillo, I. & Zwahlen, F. 2001. Evaluation and validation of vulnerability concepts using a physically based approach. 7th Conference on Limestone Hydrology and Fissured Media, Besnacon 20–22 Sep. 2001, *Sci. Tech. Envir., Mém. H.S*, 13: 67–72.

Cornaton, F. & Perrochet, P., 2001. Notes concerning the experimental program VULK. Manuscript, archive of the CHYN Université de Neuchâtel, Switzerland: p. 5.

Jeannin, P.-Y., Cornaton, F., Zwahlen, F. & Perrochet, P. 2001. VULK: a tool for intrinsic vulnerability assessment and validation. 7th Conference on Limestone Hydrology and Fissured Media, Besnacon 20–22 Sep. 2001, *Sci. Tech. Envir., Mém. H.S.*,13: 185–190.

Kullman, E. 1990. Krasovo-puklinové vody. *Karst-fissure waters*. GÚDŠ, Bratislava: p.184.

Michalko, J. & Vojtková, S. 2003. Environmentálne izotopy v podzemnej vode prameňov Periodická vyvieračka a Teplica v Tisovskom krase. *Podzemná voda* IX/2003 (1), Slovenská asociácia hydrogeológov, Bratislava: 14–22.

Padúch, F. 1998. Zaujímavosti hydrologického režimu Periodickej vyvieračky v Tisovci. *Práce a štúdie* 57, SHMÚ, Bratislava: 7–17.

Wiesengangerová, S. 2000. Hydrogeologické pomery Tisovského krasu. Dipl. práca, Manuskript – archív Katedry hydrogeológie PRIF UK, Bratislava: p. 102.

Zwahlen, F. (ed.) 2004. Vulnerability and risk mapping for the protection of carbonate (karst) aquifers. *COST Action 620 Final Report*. Office for Official Publications of the European Communities, Luxembourg, 2004 – XVIII: p. 297.

CHAPTER 21

Vulnerability of the karst – fissured Upper Jurassic aquifer of the Cracow Jurassic Region (Poland)

J. Różkowski, A. Różkowski & J. Wróbel
University of Silesia, 41-200 Sosnowiec, Poland

ABSTRACT: The Upper Jurassic karst – fissured aquifer is the main source of groundwater supply in the Cracow Jurassic Region. During the last years several regional reports of the Jurassic karst – fissured aquifer have been made in which the groundwater vulnerability has been evaluated and presented in the maps or only discussed (Różkowski 1996, 1997; Różkowski et al. 2001). Vulnerability assessment of the aquifer was calculated according to the empirical formulae taking into account seepage time of water and pollutants from the surface to the Jurassic aquifer. In this paper, groundwater vulnerability to the surface contamination evaluated in the map by the empirical formulae has been confronted with the results of field experiments of vertical seepage time through the vadose zone as well as the calculated values of the residence time of groundwater in aquifer.

1 INTRODUCTION

The Cracow Upper Jurassic Region located in the southern Poland covers the area of about 650 km². It is a southern part of the Upper Jurassic limestone Upland belt stretching from Cracow in the southeast to Wieluń in the northwest. The hydrogeologic conditions of the region are typical for karst areas such as scarity of surface water and abundance of groundwater.

The present state of the groundwater contamination oblige us to determine the strategy of the Upper Jurassic aquifer protection. For this purpose, the groundwater vulnerability maps have been constructed. The vertical seepage time for potential contaminants from the ground surface being the base for vulnerability valuation and the risk range to the Upper Jurassic aquifer were determined.

Hydrogeological environment of the Upper Jurassic karst – fissured carbonate aquifer, the concept of the aquifer vulnerability mapping and the results of the valuation made, as well as comparison of theoretical calculation results with the field tests data are shown in the following sections of this paper.

2 GEOLOGY

The Cracow Upper Jurassic Region (CUJR) consists of the Upper Jurassic carbonate sequence underlain by older Jurassic, Triassic or Paleozoic deposits (Buła 2000). The

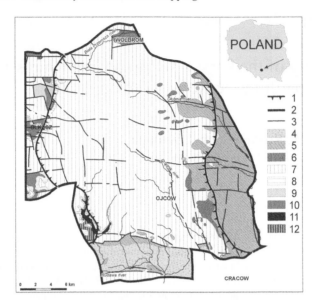

Figure 1.   Generalized geological map of the study area (excluding Quaternary deposits).
1 – boundary of the Upper Jurassic aquifer, 2 – boundary of the numeric model, 3 – main faults, 4 –
Tertiary, 5 – Upper Cretaceous, 6 – Lower Cretaceous, 7 – Upper Jurassic, 8 – Middle Jurassic, 9 –
Upper Triassic, 10 – Middle Triassic, 11 – Lower Carboniferous, 12 – Devonian.

Upper Jurassic sediments are partly covered by Cretaceous and Cenozoic deposits (Figure 1).
The thickness of the Upper Jurassic carbonate massif is variable in a range from 40 to
about 300 m. Three general types of calcareous rocks may be differentiated: (1) thin –
layered marly limestone and marl in the bottom part and in the upper part of the sequence,
(2) layered micritic, platy limestone, (3) biothermal limestone, massive and chalky limestone. Within this third type of rocks occur large caves and caverns with up to 20 m height
sinkholes and large karst depressions more than 1 km in diameter. All these forms are
filled partly or completely with Cenozoic deposits.

The Upper Jurassic carbonate massif was affected by karstification mainly during the
two periods – in the Early Cretaceous and in the Cenozoic. The karst relief and karst
groundwater paths were mostly "fossilized" during the glaciations. Since the retreat of the
Warta (penultimate) Glaciation, the exhumation of the paleokarst sculpture and of underground water paths is occurring (Głazek et al. 1992).

The Cretaceous sequence of variable thickness (from 1 up to 30 m) represented by sandy
deposits of Albian-Cenomanian age and marls of Senonian age occurs on the east slope of
the karst aquifer. The thick up to 80 m layer of Miocene clayey sediments occurs in overburden in the southern part of the region in the area of Krzeszowice Graben (Figure 1).

The Quaternary sediments, thick from 1 to 15 m, locally to 40 m, are represented by loams
and sands in river valleys and by loesses and diluvium deposits on uplands.

3   HYDROGEOLOGICAL SETTING

The Upper Jurassic aquifer is the main aquifer of the CUJR. It is carbonate karst-fissured-
porous aquifer. This aquifer is covered by Quaternary sediments, locally Tertiary and

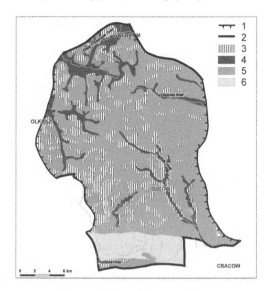

Figure 2. Lithological characteristic and permeability of the vadose zone of the Upper Jurassic aquifer.
1 – boundary of the Upper Jurassic aquifer; 2 – boundary of the numerical model; 3 – outcrops of the carbonate rocks of the Upper Jurassic ($k = 10^{-5} - 10^{-3}$ m/s); 4 – Quaternary loam and sand in river valleys on the Jurassic sediments ($k = 10^{-5} - 10^{-4}$ m/s); 5 – loess, Quaternary diluvial sediments and Cretaceous marls on the Upper Jurassic limestone ($k = 10^{-6} - 10^{-5}$ m/s); 6 – Tertiary sand, clay, loess and silt ($k = 10^{-4} - 10^{-8}$ m/s)

Cretaceous ones, of differentiated permeability (Figure 2). In the eastern part of the region, the aquifer lies under Cretaceous deposits of the Miechów syncline. Thickness of water-bearing limestone of the Upper Jurassic is differentiated and it is in the western part of the aquifer 20–50 m, in the eastern part – over 250 m, in the southern part 30–100 m and in the northern part – about 300 m. The carbonate series are strongly cracked and fractured.

In the recharge zone of the carbonate karst – fissured complex, there are three hydrogeological zones lying one upon another. These are the vadose zone with gravitational vertical flow, the phreatic zone with the lateral water flow and the intermediate zone situated between the previous ones.

The boundaries between hydrogeological zones mainly related to geomorphological factors, geologic structure and hydrodynamic conditions variable in time. The range of the vadose zone depends on the land relief and depth of the drainage base. In the investigated karst – fissured aquifer, the occurrence of the shallow phreatic zone with an active flow and the deep one with slow laminar flow typical for karstic aquifers have not been noticed. It seems to be mainly related to the low thickness of the Upper Jurassic carbonate complex.

The Upper Jurassic aquifer forms one hydraulic system. Rock matrix blocks are treated as storage elements and fissures and karstic channels separating them – as transmissivity elements (Motyka et al. 1993).

Massif of the Upper Jurassic limestone is characterized by values of hydrogeological parameters done below. Open porosity is from 0.6% to 18.9%, locally for chalky limestone up to 27.8%; geometric mean value is 4.4%. Gravity drainage capacity is from 0% to 13% and geometric mean is 0.6%. Fracture porosity of limestone within the vadose zone, according to field measurements in quarries done by J. Różkowski and A. Polonius, is from 0.03% to 2.98% what corresponds with hydraulic conductivity of fracture permeability from $2.23*10^{-5}$ to $5.05*10^{-1}$ m/s.

Values of fracture porosity within the phreatic zone are from 0.06% to 0.48% and of karstic porosity are from a fraction to 6.6% (Liszkowska 1990).

Values of the hydraulic conductivity for the phreatic zone of the karst – fissured water – bearing complex of the Upper Jurassic carbonate series calculated from the results of test pumping are variable in the range from $2.3*10^{-8}$ m/s to $6.5*10^{-3}$ m/s, while geometric mean is $1.1*10^{-4}$ m/s.

Values of discharge from 0.1 to 5 m³/h are dominant (35% of population) among 153 wells. Only 5% of studied population oversteps discharge over 50 m³/h. Maximal discharge values are observed within tectonic dislocation zones where karst is developed.

Within the upland areas, the depth of water table is from 30 m to 70 m. In deeply cut, up to 100 m, karstic canyons depth of water table occurrence is from a few meters to 30 m.

Configuration of hydrodynamic field shows the occurrence of the three types of flow within the aquifer: regional, intermediate and local ones (Różkowski Stachura 1971). Intermediate and local flows make 70% of underground runoff. The underground drainage basins of local flow are drained by streams net and the ones of intermediate flow – by river valleys of the: Biała Przemsza, Dłubnia, Prądnik, Rudawa rivers (Figure 1). Regional flow in the Upper Jurassic aquifer is in the eastern direction toward Miechów syncline.

Various values of hydraulic conductivity in particular hydraulic systems of the Upper Jurassic carbonate complex cause different water flow rate. In the channel systems, the flow rate values are in the range of tens of thousands m/year, whereas in the fissure systems, tens of hundreds to several thousands m/year. Flow through porous system of rock matrix is from decimal of meter to a dozen or so meters in a year. The average flow rate through cracked and cavernous carbonate rock complex, considering its matrix porosity, is in the range from several hundreds to over one thousand m/year.

Recharge of aquifer takes place directly or indirectly through Quaternary sediments. Karst-fissured-porous type of aquifer and the occurrence of permeable overburden favour intensive infiltration of atmospheric water and groundwater renewal. It also causes reduction of surface flow. Diffuse recharge over the whole outcrops is dominant. Concentrated recharge takes place only in a small number of swallow holes.

Epikarstic zone, up to 9 meters thick, is very fractured and karstic one. Significant differentiation of its permeability in comparison with the vadose zone lying beneath is a reason for the occurrence of suspended horizon within epikarstic zone and delay in vertical flow from a few to a dozen or so weeks (Leszkiewicz Różkowski 2000).

Numerical modelling with the application of MODFLOW program enabled an estimation of underground water resources renewal of the Upper Jurassic aquifer (Różkowski et al. 2001). Renewal of resources in the individual drainage basins is differentiated in the range from 4.85 to 6.06 dm³/s*km².

4   THE CONCEPT OF MAPPING OF THE UPPER JURASSIC AQUIFER VULNERABILITY

The concept of mapping of the Upper Jurassic aquifer vulnerability to anthropogenic pollution is based on the assumption that the vulnerability is the natural property of the water-bearing system depending on its "sensitivity" to the influence of pollution sources. Natural water vulnerability has been defined as the degree of pollution risk to the Jurassic water-bearing system in the case of the anthropogenic contaminants.

The main factors inhibiting the migration of pollutants to groundwater are regeneration processes of water seeping through soil and rocks of the vadose zone and the degree of the aquifer isolation from the ground surface. It is assumed that the longer time of water residence in the vadose zone, the more active processes of the regeneration of the polluted water. Vulnerability of the Jurassic aquifer was estimated in this paper only by the travel time of water vertical seeping from the ground surface through the vadose zone to the phreatic zone because of the lack of data concerning the soil structures and thickness.

It should be pointed out that the estimation of the potential vulnerability of the Upper Jurassic aquifer was a very difficult task due to a few only direct measurements of the water seepage time through the vadose zone. It necessitated recognition of the thickness and permeability of the overburden formations as well as the Jurassic carbonate complex in the vadose zone.

Lithological characteristic of the vadose zone is based on the geological maps of Poland in scale 1:50,000. Moreover, the geologic profiles of 50 boreholes and 176 drilled wells were taken into account.

The formulae used for "seepage time" calculation also required knowledge of the quantity of effective recharge of the Upper Jurassic aquifer. The value of average atmospheric precipitation for the studied area is 720 mm. Numerical modelling with the application of MOD-FLOW program enabled an estimation of recharge quantity of the Upper Jurassic aquifer (Różkowski et al. 2001). Quantity of recharge obtained from modelling is variable in the range from 10 to 273 mm/year. High values in range from 150 to 273 mm/year are characteristic of the upland areas (Figure 3).

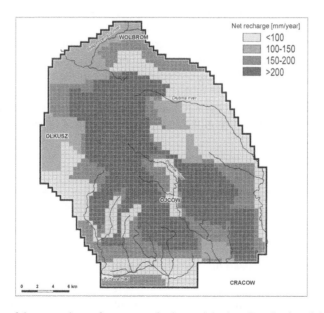

Figure 3.   Map of the net recharge from atmospheric precipitation. Result of model calibration.

In case when the hydrogeological profile of the vadose zone was well known the seepage time (t) has been calculated using the N. N. Bindeman's simple formula, describing the vertical flow rate through the mentioned zone:

$$t_a = \sum_{i=1}^{n} m_i/(v_a)_i \qquad v = 1/n_o \sqrt[3]{\omega^2 k} \qquad (1)$$

where: t_a – seepage time (year)
$\quad v_a$ – seepage velocity (m/year)
$\quad m_i$ – bed thickness (m)
$\quad k$ – hydraulic conductivity (m/year)
$\quad n_o$ – effective porosity
$\quad \varpi$ – infiltration intensity (m/year)

The water seepage time through the vadose zone in the area of the analyzed aquifer has been defined on the base of geologic profile, estimated infiltration rate and assumed values of volume moisture of rocks.

The following S. Witczak and A. Żurek (2002) modified formula has been applied in the calculations:

$$t = (1000*w*m)/ \varpi \qquad (2)$$

where: t – seepage time (year)
$\quad w$ – rock moisture volume
$\quad m$ – thickness of isolation cover (m)
$\quad \varpi$ – infiltration intensity (mm/year)

The calculated average time of the vertical seepage through the vadose zone with variable infiltration intensity is shown in the Table 1.

With a view to calculating the time of the vertical seepage through the vadose zone in the investigated area, the scheme of the geologic conditions of the vadose zone has been made. For this purpose the maps and geologic-structural profiles of the roof and the base of the Jurassic carbonate complex, the map of the Quaternary formation thickness and occurrence of clay layers, with the thickness of more than 5 m and the contour maps of the carbonate complex and the topographic map of the area have been applied (Różkowski et al. 2001).

Table 1.   Average time of the vertical seepage through 1 m of the vadose zone.

No	Lithology of the vadose zone	Infiltration intensity (mm/year)	Rock moisture volume	Time of water seepage through 1 m of the vadose zone (year)
1	Sands	200–250	0.09	0.36–0.45
2	Loesses	100–120	0.30	2.5–3.0
3	Sands and tills	80	0.30	3.75
4	Marls	100–150	0.15	1.0–1.5
5	Limestones	150–273	0.02	0.07–0.13

Thickness of the vadose zone is defined by the hydroisobath maps of the Upper Jurassic aquifer (Różkowski 1997).

To assess the time of the vertical seepage through the following, lithologically variable strata of the covering rocks, the previously presented empirical formula (2) has been applied which, after modification and adaptation to the multi-layer profile, has the following form:

$$t_v = [(m_1*w_1) + (m_2*w_2)*(m_3*w_3) + (m_4*w_4) + (m_5*w_5)]*1000/W \tag{3}$$

where: t_v – time of vertical seepage to the phreatic zone (years)

   W – infiltration intensity of atmospheric precipitation (mm/year)

   m_{1-5} – thickness of the succeeding lithologic layers shown in the table 1 (m)

   w_{1-5} – rock moisture volume shown in the Table 1

Calculation of the time of the vertical seepage according to the presented formula has been made in the square net applied in the numeric modelling (Różkowski et al. 2001). This enabled oneself to use in calculation the distribution of effective infiltration received from the aquifer model calibration.

As a result of the calculation, the values of the vertical seepage time in the range from less than 1 year to 100 years have been found. The calculated time values are related to natural vulnerability, meaning seepage of the conservative chemical substances from the ground surface.

Groundwater vulnerability to pollution evaluated in the areal map by the empirical formulae (Figure 4) has been confronted with the results of direct point field tests.

The Upper Jurassic aquifer vulnerability has been evaluated indirectly by residence time of groundwater in it. The results of assessment of tritium isotope concentrations in water have been used (Różkowski 1996).

Considering the calculated time of the vertical seepage through the vadose zone, five classes of aquifers vulnerability have been distinguished:

– very high – seepage time less than 2 years
– high – seepage time from 2–5 years
– medium – seepage time from 5–25 years
– low – seepage time from 25–100 years
– very low – seepage time more than 100 years.

5   RESULTS OF ASSESSMENT OF GROUNDWATER VULNERABILITY

Evaluation of vulnerability of the Upper Jurassic aquifer has been based upon the analysis of the time rate of vertical seepage of water from the surface through the vadose zone. Taking into account the results of evaluation, the Jurassic aquifer in Cracow region is of various vulnerability. For the limestone outcrops apparent in the area of the Cracow Upland, the time of vertical seepage usually does not exceed 1.6 years, whereas in the area covered by Quaternary and Cretaceous overburden, it ranges from 3.0–100 years, mainly 20 years (Figure 4). The aquifer is naturally protected by the clayey Tertiary sediments cover only in a small area within the Krzeszowice Graben (Figure 1). Taking into account these figures a very high and high vulnerability is typical for the outcrops area of the

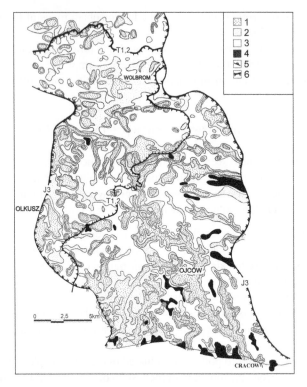

Figure 4.   Map of vulnerability of the Upper Jurassic fissure-karst aquifer.
Vulnerability classes: 1 – extremely high ($t_v < 2$ years), 2 – high (t_v = 2–5 years), 3 – medium (t_v = 5–25 years), 4 – low (t_v = 25–100 years), 5 – boundary of the fissure-karstic Jurassic aquifer, 6 – extent of the underlying Triassic aquifer.

upland, constituting 57% of the total area under investigation (Figure 4). The zone of Jurassic strata under the thin cover of Cretaceous marls or Quaternary loesses as well as slope wash sediments is of medium and low vulnerability. These sediments cover the Jurassic limestone on the 43% of the area.

These figures are confirmed by the results of a few quantitative analyses of tritium content in groundwater. The tritium investigations of the local flow system carried out by J. Różkowski (1996) have shown that the groundwater circulation time in the karst – fissured aquifer can be about four – seven years. The point tritium measurements in the upper most part of the upland have demonstrated that the seepage time of atmospheric water through 16 m of Quaternary sand and loam to the karst aquifer can be less as ten years.

Groundwater vulnerability of the Upper Jurassic aquifer evaluated in the areal map by the empirical formulae (Figure 4) can be confronted with the results of a few only field experiments. Conservative contaminant transport through the vadose zone of the Upper Jurassic aquifer was tested by hydrogeologists of the University of Science and Technology in Cracow (Kleczkowski, ed. 1999).

The field experiments of vertical flow through the vadose zone were carried out in the years from 1996 to 1999 in the vicinity of Cracow. Chloride ion was the most often used tracer. The vadose zone of the Jurassic rocky limestone was penetrated to the depth of 15–72 m. Velocity of vertical flow through the rocky limestone in the uncovered horst areas varies from 146 to 3650 m/year. It means that time of water seepage through 1 m of the vadose zone was shorter than 0.007 of year. There is a lack of measurements of seepage time through covered limestone.

The field experiments of vertical flow through the sandy layers indicate flow velocity from 5.5 to 13 m/year, through sandy tills – 0.2 m/year and in the case of loesses 0.34 m/year. Time of flow through 1 m of these deposits was respectively: 0.18, 5.0 and 2.9 of year. The laboratory experiments in filtration column revealed that the transport time through the tested clayey–sandy–silty formation of the Quaternary cover was about 0.2 m/year. It means that time of flow through 1 m takes time about 5 years. We can assume that the figures from the field experiments are close to those obtained from the theoretical calculations (Table 1). It does not work in the case of uncovered limestone of the horst areas.

Comparison of the theoretical calculation results with the field tests data indicate to the possibility of using the theoretical formulae for general assessment of the groundwater vulnerability. The accumulated experience, however, does render it necessary to carry out further complete investigations for drawing up the more detailed groundwater vulnerability maps.

REFERENCES

Buła, Z. 2000. The Lower Paleozoic of Upper Silesia and West Małopolska. *Prace PIG CLXXI.* Warszawa.

Glazek, J., Pacholewski, A. & Różkowski, A. 1992. Karst aquifer of the Cracow-Wielun Upland, Poland. *Int. Contributions to Hydrogeology*, vol. 13. Verlag Heise, Hannover: 289–306.

Kleczkowski, A.S. (ed.). 1999. The velocity of contaminant transport through unsaturated zone from field and laboratory experiments. Academy of Mining and Metallurgy, Cracow.

Leszkiewicz, J. & Różkowski, J. 2000. Response of karstic – fissured springs to infiltration recharge in the area of Ojców National Park (Cracow Upland, southern Poland). *Kras i Speleologia*, 10 (XIX). Wyd. Uniw. Ślaskiego. Katowice: 27–44.

Liszkowska, E. 1990. Badania szczelinowatości dla potrzeb określenia modelu geohydraulicznego i migracyjnego ośrodków szczelinowych i szczelinowo – krasowych. W: Szczelinowo – krasowe zbiorniki wód podziemnych Monokliny Śląsko – Krakowskiej i problemy ich ochrony. CPBP 04.10. Ochrona i kształtowanie środowiska przyrodniczego. Wyd. SGGW-AR. Warszawa: 73–83.

Motyka, J., Pulido – Bosh, A. & Pulina, M. 1993. Wybrane problemy hydrologii i hydrogeologii krasowej w skałach węglanowych. *Kras i Speleologia*, 7 (XVI). Wyd. Uniw. Śląskiego. Katowice: 7–19.

Różkowski, A. & Stachura, A. 1971. Warunki filtracji wód w utworach jury na przykładzie wybranego obszaru w południowej Polsce. *Kwart. geol.* T.15 nr 3: 671–687.

Różkowski, J. 1996. Transformations in chemical composition of karst water in the southern part of the Cracow Upland (Rudawa and Prądnik drainage areas). *Kras i Speleologia*, nr specjalny 1 (1996). Wyd. Uniw. Śląskiego. Katowice.

Różkowski, J. 1997. Map of occurrence, usage, vulnerability and protection of the Upper Jurassic Major Groundwater Basin. In: Map of occurrence, usage, vulnerability and protection of fresh – groundwaters in the Upper Silesian Coal Basin and its margin (Różkowski A., Rudzińska – Zapaśnik T. & Siemiński A. ed.) 1:100,000. PIG. Warszawa.

Różkowski, J., Kowalczyk, A., Rubin, K. & Wróbel, J. 2001. Renewal of fissure – karstic groundwater in the Upper Jurassic aquifer in the Cracow Upland area on the base of modelling study. *Współczesne problemy hydrogeologii*, X. Wrocław-Krzyżowa: 245–252.

Witczak, S. & Żurek, A. 2002. Vulnerability assessment in fissured aquifers. In: *Groundwater quality and vulnerability*. Silesian University. Sosnowiec: 241–254.

CHAPTER 22

Intrinsic vulnerability assessment for the Apulian aquifer near Brindisi (ITALY)

M. Spizzico, N. Lopez & D. Sciannamblo
Department of Civil and Environmental Engineering Polytechnic of Bari, Bari, Italy

ABSTRACT: The evaluation of the intrinsic vulnerability in a part of the Apulian deep aquifer has been investigated on a wide stretch of the Adriatic coast, in the province of Brindisi (Italy). The Apulian aquifer is deep, fissured, karstified, and coastal in type. Groundwater flows over intruding seawater. The aquifer is largely homogeneous, but small scale features may vary. In the area under investigation, the main anthropic impact on the aquifer is represented by an industrial plant close to Brindisi town. There is also significant impact on groundwater caused by the intensive agricultural practice in the area. In order to estimate the intrinsic vulnerability of the Apulian deep aquifer, a parameter-system based on scores and weights (PCSM: Point Count System Models) already used in other Italian regions, has been used.

The model adopted, SINTACS[1], by considering geomorphologic features, the soil nature, the over-burden geometry, the structure of the hydrogeological system, and recharge process, allows potential receiving groundwater to be located and water-carried contamination to be spread and lessened. This work is intended to define the model applicability to the vulnerability evaluation of a portion of the Apulian carbonate coastal aquifer.

1   MAIN FEATURES OF INVESTIGATED ZONE

The study area is about $340\,km^2$, situated between the cities of S. Vito dei Normanni and Brindisi in Italy. It is a coastal area with the carbonate plateau of Murge extending west and a wide plain formed by terraced sea deposits, in the east. Topographical varies up to 135 m a.s.l. and slopes generally are between 0 and 2%, reaching max values of 4–5% in a limited zone of eastern portion.

In the northern part, behind the coast, there is a low-lying and swampy zone (the wet land of Torre Guaceto). Runoff to the sea mainly takes place by means of natural rills, partially canalized near the coast, among which the main one "Canale Reale" which also receives waste water from towns and farms along its course. The area under study belongs to the so-called Apulian Platform, basically made up of calcareous and calcareous–dolomitic rocks, fractured and karstified with to variable degrees, and outcropping in the south-western zone (Figure 1).

---

[1]SINTACS is acronym of **S**oggiacenza (Depth to water), **I**nfiltrazione efficace (Infiltration), **N**on saturo (Unsatured), **T**ipologia di copertura (Type of soil), **A**cquifero (Aquifer), **C**onducibilità idraulica (Hydraulic conductivity) e **S**uperficie topografica (Topographical surface).

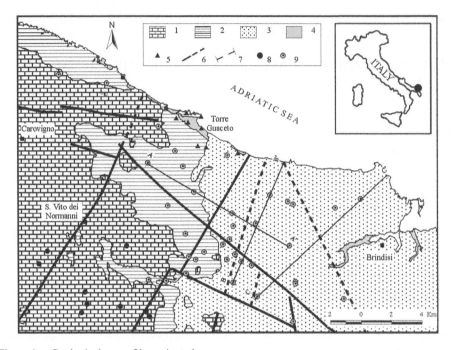

Figure 1.   Geological map of investigated area.
1 – limestones (Cretaceous); 2 – calcarenites (lower Pleistocene); 3 – sands with interbedding of sandstones and marly clays (middle-upper Pleistocene); 4 – recent and actual deposits (Holocene); 5 – spring; 6 – fault; 7 – cross-sections; 8 – sink-hole; 9 – borehole.

The "deep aquifer" occurs in the carbonates rocks floating above intruding seawater, and flowing to the Adriatic Sea under low piezometric gradients (Grassi & Tadolini 1985; Zezza 1978).

Calcarenites of Calabrian age (Figure 2) represent the base of the "Sedimentary cycle of the Fossa Bradanica" deposited along the eastern border of the foredeep basin (Ciaranfi et al. 1992). They are of variable degrees of cementation, with thickness not exceeding 20–30 metres. There is a middle-upper Pleistocene composed of yellow sands and a base level of marly clays (Figure 2). These deposits are found in the south-eastern part of the zone with maximum thickness of 40–50 metres, containing scarce groundwater of seasonal nature.

The area has regional structural features (Area of Murgia-Salento). The carbonate basement is displaced in blocks, locally degrading towards the Adriatic Sea and reaching –40 metres b.s.l. The main system trends NW-SE and E-W; the secondary system is represented by structures oriented SW-NE. Near the coast the marly clays inhibit free discharge of deep aquifer water to the sea. In a limited area, where the calcarenite deposits crop out, groundwater circulation is phreatic, whereas in the western zone, because of more compact limestone, groundwater is confined (Tadolini et al. 1994). The permeability of the limestone and calcarenite formations is variable. In a earlier work (Sciannamblo et al. 1992), one can notice that there is an alternation between higher permeable rocks and lower ones, according to strips almost parallel and perpendicular to the coast-line.

The deep aquifer, fed by rainwater occurring in the innermost upland plain of the Murgia, discharges to the sea by means of three main routes. Two of them converge in the

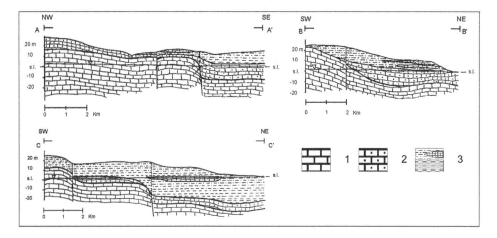

Figure 2.   Geological cross-sections.
1 – limestones (Cretaceous); 2 – calcarenites (lower Pleistocene); 3 – sands with interbedding of sandstones and marly clays (middle-upper Pleistocene).

coast as sub-aerial and submarine springs. The third one is oriented towards the marsh area, where waters riseflow freely towards the sea. Coastal discharge, at sites well localised, on the average is 300 l/s; the same discharge is for natural depressions in proximity of the sea.

3   ELEMENTS OF ANALYSIS TO DRAW UP A VULNERABILITY MAP

The territory under investigation has been subdivided by means of Square Finished Elements (SFE) represented by square meshes with 500 m. sides. Each of the 1374 SFEs has been provided with a special form summarising all known information. Each form, besides a weighted average stratigraphy, includes information on the hydraulics and soil type. Each cell was later attributed a "score" in relation to each parameter considered in the model (Table 1) and a "weight" according to an acknowledged potential impact on territory.

3.1   *Depth to water (S)*

In the zone where groundwater flows freely (Figure 3 and 4) depth to water was calculated for each cell for the period March–April 2002 from 90 wells.

3.2   *Infiltration (I)*

The effective infiltration has been calculated by taking into account climatic, lithologic, and pedologic factors typical of each cell. Inter-annual average rainfall (P) has been obtained by reconstructing the thirty-year data series collected from three stations of the Italian Hydrographic Service, (Brindisi, Latiano, e Ostuni). Evapotranspiration (Evtp) was determined, by traditional methods, for each of the three Thiessen polygons, and then adjusted to take into account the different types of vegetation. Previous studies on the Apulian territory prove that evapotranspiration represents from 20% to 40% of rainfall.

Table 1.   Scores assigned to each parameter of model.

Depth to water		Type of cover	
Range (m)	Score	Soil texture	Score
0–1	10	Thin or missing soil	10
1–4	9	Sand	8
4–6	8	Sandy loam	6
6–8	7	Loam	5
8–10	6	Clay loam	3
10–20	5	*Aquifer*	
>20	4		
		Type	Score
Infiltration		Karstified limestones	9
Lithology	χ	Fractured limestones	8
Karstified and fractured limestones	0.9	Little fractured limestones	7
Sands	0.8	*Hydraulic conductivity*	
Calcarenites	0.5		
Soil texture	χ	Range (m/s)	Score
Sandy loam, sand	0.4	$5*10^{-3} - 10^{-2}$	10
Loam	0.2	$10^{-3} - 5*10^{-3}$	9
Clay loam	0.05	$5*10^{-4} - 10^{-3}$	8
		$10^{-4} - 5*10^{-4}$	7
Active recharge (mm/y)	Score	$5 - 10^{-5} - 10^{-4}$	6
		$10^{-5} - 5*10^{-5}$	5
250–325	9	$5*10^{-6} - 10^{-5}$	4
175–250	8		
150–175	7	*Topographic surface*	
125–150; 400–450	6		
100–125; 450–500	5	Slope classes (%)	Score
75–100; >500	4		
60–75	3	0–2	10
<50	1	2–4	9
		4–6	8
Unsatured zone			
Lithology	Score		
Karstified limestones	9–10		
Fractured limestones	7–8		
Little fractured limestones	6		
Calcarenites	5		
Sandy clay	4		
Clay	2		

Finally, the infiltration coefficient (χ) (Civita & De Maio 1997) was taken from tables of the model, by considering local lithologic and pedologic characteristics.

3.3   *Unsaturated zone (N)*

Information related to unsaturated zone, that is the whole lithological layer between the lower boundary of top soil or, if any, between ground level and depth where groundwater is located, was obtained from stratigraphic surveys. The zone is characterised by four main lithological types: variously fissured and karstified limestone, calcarenite, sandy clay and clay. Each cell was scored by means of weighted average in relation thickness.

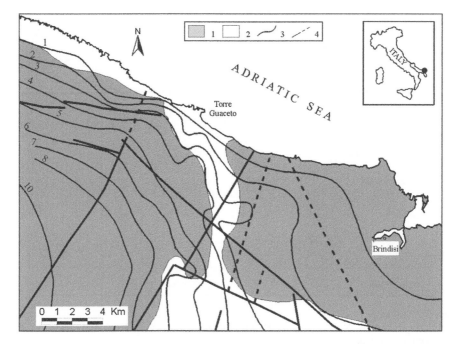

Figure 3.   Piezometric and hydraulic circulation map.
1 – circulation under confined condition; 2 – circulation of freatic type; 3 – piezometric surface;
4 – fault.

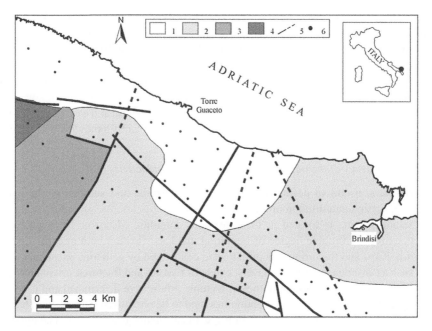

Figure 4.   Depth to water map (from topographic surface).
1 – 0–50 m; 2 – 51–100 m; 3 – 101–200 m; 4 – >200 m; 5 – fault; 6 – well.

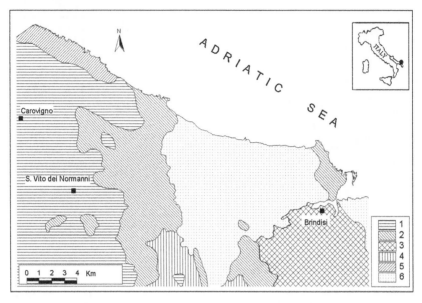

Figure 5.   Soil map.
1 – thin, clay loam; 2 – deep, clay loam; 3 – very deep, loam; 4 – very deep, sandy loam; 5 – thin, sandy loam; 6 – very deep, sandy.

3.4   *Type of soil (T)*

Information concerning soil type was obtained from the pedological Map of the Apulian Region (2001) and integrated with analytical data collected during an investigation of soil agricultural characterisation in the province of Brindisi (Lopez 1971). Areas lacking in data were investigated directly (Figure 5).

In general, the western area, where limestone and calcarenite outcrop, is characterised by a thin and discontinuous cover (ruptic and lithic soils), except for some limited zones near small dolines where the thickness can exceed 1 metre. In the central and eastern sectors near the Brindisi, soils are thick and very thick and they are dominantly sandy loam or sand.

3.5   *Characteristics of the aquifer (A) and hydraulic conductivity (C)*

The parameters, useful to define the vulnerability, describe and summarise the hydro-dynamic dispersion and dilution processes.

The saturated zone is located in the calcareous-dolomitic Mesozoic basement which shows an extremely variable degree of fracturing and karstification both in expansion and deep down. Karst and fracturing was investigated cell by cell by gathering subsidiary information such as: average thickness of layers, extent of fissures and fractures, orientation, position and interconnection. Discontinuities of tectonic origin were determined and classified according to their age. The older discontinuities tend to be obstructed with calcium hardening or fine-grained residual fills, and, at times, with calcareous clastic rocks. The hydraulic conductivity for the area was obtained by determining "relative specific flow" resulting from numerous pumping tests on more than 80 wells using the Babouchkine-Guirinsky's

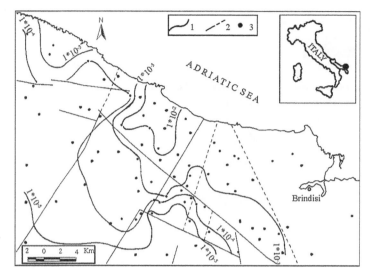

Figure 6.   Hydraulic conductivity map (m/s).
1 – hydraulic conductivity; 2 – fault; 3 – well.

expressions (Castany 1963). Hydraulic conductivity ranges from $1*10^{-6}$ m/s to $4*10^{-2}$ m/s. (Figure 6).

3.6   Topographical surface (S)

Information related to this parameter were obtained by using a DTM (Digital Terrain Model) with 100 m resolution and resampled to 500 m to obtain average slope values for each cell. Three slope classes were determined and given a score (Table 1).

3.7   Weight strings

Five weight strings were applied, each indicating a territorial condition of potential average impact ("normal impact"), diffuse source contamination ("great impact") and the aquifer recharge possibility, though seasonal only, in reaches of channels where water flows occur ("drainage"). "Fissured string" was mainly applied in zones with no soil and where unkarstified limestone outcrops, or where karst is only superficial. In wide karstified areas, both superficially and deep down (coastal zones) were attributed to the "karstism" string.

For cells requiring more weight strings, the one best corresponding to the most serious impact was adopted. Each cell was given scores (P_j) and weight strings (W_j) and the intrinsic vulnerability was calculated, using this formula:

$$I_{SINTACS} = \sum_{j=1}^{7} Pj*W_j$$

The results are showed in the vulnerability map (Figure 7).

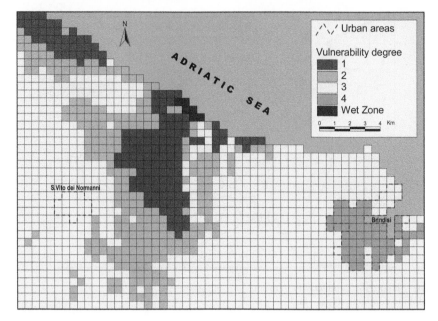

Figure 7.   Vulnerability map.
1 – extremely high; 2 – very high; 3 – high; 4 – medium.
(In the "Torre Guaceto" wet zone, where deep groundwater emerges, the model is not applied.)

4   CONCLUSIONS

The work has made it possible to evaluate the vulnerability of a large coastal zone of the carbonate. The high vulnerability dictates a high degree of prevention measures to be undertaken, which a more comprehensive study and management of the territory and agricultural activity.

REFERENCES

Castany, G. 1963. Trattè pratique des eaux souterraines. Ed. Dunod, Paris.
Ciaranfi, N., Pieri, P., & Ricchetti, G. 1992. Note alla carta geologica delle Murge e del Salento (Puglia centromeridionale). *Estr. Mem. Soc. Geol. It.*, 41. Roma.: 449–460.
Civita, M. & De Maio, M. 1997. SINTACS – Un sistema parametrico per la valutazione e la cartografia della vulnerabilità degli acquiferi all'inquinamento. Metodologia e automatizzazione. Pitagora Ed., Bologna.
Grassi, D. & Tadolini, T. 1985. Hydrogeology of the mesozoic carbonate platform of Apulia (South Italy) and the reasons for its different aspects. *International Symposium on karst water resources.* Ankara: 293–306.
Lopez G. 1971. Studio dei terreni agrari della provincia di Brindisi. *Annali Ist. Sper. Agr.* 2. Bari.
Regione Puglia 1994. Piano regionale di risanamento delle acque.
Regione Puglia 2001. Sistema informativo dei suoli della Regione Puglia, in scala 1:50000. *Carta pedologica progetti ACLA I, ACLA II e INTERREG II Italia-Albania.*
Sciannamblo, D., Spizzico, M., Tadolini, T. & Tinelli, R. 1992. Lineamenti idrogeologici della zona umida di Torre Guaceto (Br). *Geologica Romana*, 30: 754–760.
Tadolini T., Spizzico M. & Sciannamblo, D. 1994. Time course of radon concentration in the coastal belt North-East of S. Vito dei Normanni (Brindisi, ITALY). *Int. Symp. 13°SWIM*: 155–162.
Zezza, F. 1978. Lithological properties and geological conditions of carbonatic platform deposits related to karstic groundwater circulation in Southern Italy. *Geol. Appl. ed Idrogeol.* 13. Bari: 393–416.

Subject index

Author index

SERIES IAH-Selected Papers

Printed and bound by CPI Group (UK) Ltd, Croydon, CR0 4YY

23/10/2024

01778259-0002